Geologic Odyssey
A Journey Through Earth Science

by
Marvin L. Schroeder

Geologic Odyssey — A Journey Through Earth Science

All photographs taken by author.

Cover design by Craig Pritchett.

Library of Congress Cataloging-in-publication Data has been applied for.

ISBN 978-1491003206

PREFACE

"YOU ARE A DEAD MAN WALKING," the doctor said as he held his charts. "Your left main artery coming from your heart is 99% blocked and if it blocks completely, you're dead! You could have dropped dead anytime in the last six months!" The realization suddenly hit me that this whole thing could be quite serious as I now listened to the heart surgeon explain what needed to be done — three heart bypasses and an aortic valve replacement. A vein would also be taken from my left leg and used in the operation. "In addition," he said, "in cases like yours, we usually use a balloon pump to relieve the stress on the heart due to the anesthesia as we start the operation." Lord, I thought. How bad am I?

"What day do you think it is?" asked my wife softly as she bent over my bed in the Intensive Care Unit. I had just regained consciousness after the operation and I thought that was an odd question to ask. My wife then told me that the operation had been fully successful but that my lungs refused to cooperate so I had been kept sedated for ten days. The medical staff kept me in the Intensive Care Unit for another four days before moving me into a hospital room for four more days. At that time a medical evaluation would send me to a skilled nursing facility for twelve days where I would learn to walk again!

The hospital door to the room squeaked slightly and dim light filtered into the interior. The figure of the male nurse loomed above me as I pulled the covers over my head. He stood there and finally said, "I have been thinking about your stories and you absolutely have to write your memoirs. Your stories would make very good reading." He paused even longer. "I would like to get a copy." I glanced at my watch — it was two o'clock in the morning.

I guess I had to agree about the memoirs. I had been beguiling him with stories about the Wild West County near Jackson Hole, Wyoming, and some experiences in Alaska and the Nevada Test Site. Every time he showed up in the early mornings I felt compelled to tell him another story. However, I had to be mindful of the old adage about Mark Twain. He had been quizzed by a reporter about the many stories he had told about his life experiences. Twain had paused for a moment and said, "Well I have told many stories about my life — some of them were actually true!"

The suggestion that I write my memoirs was very appealing to me. No book had ever been published in the United States about the life, times and experiences of a field geologist. I had participated in the last great geologic survey of the American West in the 1960s and 1970s. The days of a field geologist are almost gone as that type of work is no longer considered a high priority by federal agencies. Most non-scientists would consider the earth sciences as dealing only with rocks, fossils, and dinosaurs. I regarded a book as an opportunity to show that there was much more involved in that field of study. Archaeology has dominated the movies and headlines in the recent past with Harrison Ford running amuck on the silver screen playing the role of Indiana Jones, an archaeologist involved in a number of adventures.

Certainly, I had been involved in a much more exciting life than I had anticipated when I was growing up in a rural environment in Wisconsin. Reading adventure books in a one-room country grade school when I was young was one way I expanded my world beyond the walls that seemed to close in on me. My professional life included tramping through Central Alaska on a military mission designed to negate a possible Russian invasion over the northern polar region. My brief sojourn on the Nevada Test Site was significant in that it showed the U.S. Government was hiding information about the extensive spread of radioactive clouds across the northern portion of the United States during the nuclear testing phase of the 1950s and 1960s. I was also one of the few geologists in the world to climb into a subsurface nuclear blast hole 1,400 feet below the surface, that occurring in an area south of Carlsbad, New Mexico, in Project Gnome.

These minor adventures paled in comparison with my main project which involved the geologic mapping of complex structural areas in the Jackson Hole and Grand Teton areas of Wyoming. The 6,000 miles that I rode on horseback would bring forth many interesting stories of that part of my life. This field work was in the land of the famous mountain men of the early 1800s who were engaged in exploration and the fur trade in that part of the country. However, it was not the actual field experiences that I had been involved in, but the disappointment of seeing politics and political ineptness gradually taking over part of the geological sciences that made it imperative that I write "my story."

In addition, I also had to admit it was a chance to write a non-technical book with some aspects of geology thrown in. This type of book would be valuable in enabling individuals who had only a limited exposure to earth science to obtain a greater grasp of that science. Most books on introductory geology tend to be too technical and boring which makes the reader rapidly lose interest. The life I was going to describe was anything but boring. In today's world, the fields of study in the geologic profession have grown so specialized that a geologist has a hard time communicating with another geologist who has a different field of study. Even a common rock is no longer considered a "common rock," but must be described in such flowery language and detail that the reader might want to believe that the rock came from another planet in an alien warship.

However, in the long run, there was even a more compelling reason for the memoirs, and that was an opportunity to briefly comment on the conflict between the earth sciences and the Christian religion. Ever since I had majored in geology in the 1950s, I had become more aware with each passing year of the divide that exists between the geological concept of creation and that expressed in the Bible. I have had ministers of the Christian faith turn their back on me and walk away when they learned I was a geologist. Yet, it is the geologist who has a greater concept of creation than do theologians! It is the geologist who has walked planet earth and examined its history. He has found no evidence for the seven days of creation or the world-wide biblical flood. The question then becomes, how do you reconcile the difference?

My interest in earth history also extended to the origin of our universe due to the long hours I sat around camp-fires observing the heavens. Einstein's initial concept of the universe was that of a "static" universe, one with an unexplainable outward force acting against gravity that would prevent the universe from collapsing. This theory was eventually discounted by modern science, even by Einstein himself. Hubble's concept in the 1930s of an expanding universe has continued into today's world. In fact, Einstein himself called his universe concept the "greatest and biggest" blunder of his professional career!

Yet, I had developed a theory and concept that indicated that Einstein was basically correct — more right than modern science which believes that the universe expands forever into oblivion. In my world, the universe has reached the limit of its expansion, and really is nothing more than a gigantic, rapidly spinning galaxy. God still lives — the universe is safe!

. . . .

To a younger geologist who wanted to work in God's country, it was very difficult for him to understand the complexity of the Jackson Hole area where I was working. An eastern moving structural plate collided with the old continental shield that formed the core of the United States. As this western plate tried to dive eastward underneath this old continental block, huge segments of sedimentary rock layers peeled off, stacked up like cordwood against each other. The total crustal shortening is estimated to be in excess of 15 miles. Each structural wedge came from a different depositional environment, and in some cases, the same rock formation had a very different lithology. In areas of thick cover and concealment, a geologist might have only the rock float in the soil to indicate the presence of a particular rock formation.

The structural complexity and limited exposures in that area caused many different interpretations by different geologists in previous years. Added to the complexity caused by the faulting and folding of the rock units was the great relief throughout the area which could reach a 4,000-foot difference in elevation from stream bottom to mountain top. The only way to efficiently work the area was on horseback as road access was very limited. It was my job to sort through all the interpretations as I did my field work and come up with a final geological map of the area; some geologists would always be unhappy! This way of life would end in the late 1970s. My colleagues and I had been engaged in one of the greatest geologic mapping surveys ever conducted in the American West.

The underlying theme of my professional career was the realization that nothing is ever free from politics. This book would also give me the opportunity to show how politics started affecting the earth sciences. Sometimes the political ramifications cannot be immediately realized — and often won't be until many years later. What is interesting is that politics would affect the scientific landscape no matter which political party was in power.

Perhaps the greatest affect of politics on science in the late 1950s and early 1960s was the denial by the U.S.

Government of the potential damage that had been created by the nuclear testing program. The venting of many nuclear tests to the atmosphere was not readily acknowledged by the U.S. Government. The fact that clouds carrying radioactive material would follow a more or less specific storm track across the northern United States was very significant. Local rains would cause much higher concentrations of radioactivity in certain areas. Assertions by nuclear regulators that tests would be conducted when atmospheric conditions were favorable really mean nothing. No known attempt has ever been made by the U.S. Government to ascertain whether cancer occurrences were higher along these storm tracks when correlated with local weather conditions.

Unfortunately, the political world effect on science has extended into the 21st century as will be shown by this book. Bad land-use planning on federal lands will cause environmental nightmares; even the slaughter of wild horses is politically driven. The fond memories of my career are somewhat tempered by the political ineptness that I encountered in the latter part of my career.

. . . .

The best way to read this book is to imagine yourself sitting around a campfire in Jackson Hole, Wyoming, in the shadow of the Grand Tetons. You have enjoyed a dinner of freshly caught rainbow trout from a nearby stream. The hour is growing late and the stars are starting to appear in the western sky. The coals in the campfire have started to burn down, and you huddle ever closer to the campfire, even getting an extra blanket to put over your shoulders.

There is utter silence except for the crackling of the fire as the cold air slowly creeps in without any hesitation. The Milky Way suddenly explodes above you in all its glory, with the stars seemingly to dance to the delight of the heavens. The old geologist very slowly and carefully drags his chair closer to the campfire and the stories begin. Suddenly, the dawn rays of the morning sun start streaking across the eastern horizon; you have been so immersed in the stories that you have stayed up all night. You now realize why one of the other listeners had leaned across to you as the stories began last night and said softly, "Welcome to the world of Mark Twain."

*This book is dedicated to all book readers
who have the spirit of adventure.*

The following stories will start to form an odyssey —
a journey through time and space. It is my story. A story
that needs to be told before memories fade and the pen
runs dry. This book is written for your enjoyment. So
whether you treasure it or not, there is one thing
we can always agree on, *Somebody Up There Likes Me*.
I can truly say that I would not have traded my life
for any other time in history.

ACKNOWLEDGMENTS

Every author has to acknowledge those who have been instrumental in getting him to write his best. I have to especially thank my wife Linda for her encouragement in my endeavors. My daughter Ella has helped in the preparation of several of the chapters as well as Patti Roberts who graciously typed several of the initial chapters from my handwritten notes. I owe many thanks to my neighbor and good friend, Georgiann Crouse for her efforts in reviewing and editing some of my initial chapters. In particular I want to thank Irene Gengler for being an avid cheerleader. I need to thank Chanda Ruminer for her word processing of the initial chapters. The dedication shown by Jennifer N. Herrera in the preparation of the manuscript is greatly appreciated. I owe my deepest gratitude to her as this book would not have been possible without her constant efforts to ensure the quality of the original manuscript. I owe special thanks to Craig Pritchett for his excellent design of the book cover. In addition, there are many others I should thank for their helpful comments and suggestions, but this becomes impossible when space becomes an issue. Nevertheless, thanks to all of you.

TABLE OF CONTENTS

PART I — A Geologic Odyssey

Chapter

1	Introduction	1
2	Alaskan Days	9
	A Russian Invasion must be stopped	
3	The Odyssey Begins	29
4	The Blonde from Muscle Beach	43
	I came back to Denver full of wisdom—	
	and not a dime in my pocket	
5	Return to Swan Valley	53
6	The Odyssey Continues	59
7	The Lopsided Deer of Teton County	69
	My uncle has seen them many times	
8	Jackson Hole Days	75
	The sword of Damocles hangs by a thread	
	over Jackson	
9	Horses—Horses—Horses	97
10	A Russian Spy	101
11	An Evening out for the Jackrabbits	111
12	Politics Win—Science Loses	115
	$20 Million Coal-Lode Maps Useless	
13	Sleeping with the Enemy	125
	We don't need scientists—All we need are	
	common, ordinary, run of the mill bureaucrats	
14	A Town that Committed Suicide	143
	The Town of Centralia lighted a match—	
	and the town crumbled	

15   The "Phantom" Wild Horse............ 147
      Why angels want to weep

16   The Universe—A Massive Spinning Galaxy... 155
      Time and space—was Einstein wrong?
      You too can journey to the edge of the universe

PART II — A Time For Reflections

17   The Early Years 163

PART III — Philosophy

18   Life After Death—A Scientific Explanation.....201
      I tell you, I saw an angel

19   Talking Snakes—A Christian Dilemma........... 211

Epilogue... 223
Is Washington, D.C. Ready for a Big Earthquake?

Bibliography.. 227

General Geologic References......................... 233

Non-Geologic References 237

Appendix 239

Geologic Plates..................................... 247

Appendix 2 253

Geologists Who Worked in Mapping Program.............. 266

Chapter 1

INTRODUCTION

I entered the University of Wisconsin as a freshman in the fall of 1954 after graduation from high school and declared my major to be in geology. I had given no consideration as to what employment opportunities might exist when I graduated in four years. When I obtained my Bachelor of Science degree in June of 1958, job opportunities for geologists were almost nonexistent. Over the years I have had many men tell me, "I could see the handwriting on the wall back then—there were no jobs for a geologist so I went into engineering!"

With job opportunities so bleak, I finally decided to go to graduate school. It was general knowledge among geology students that in order to get a job offer it was necessary to have an advanced degree. In many ways graduate school was one way of killing time but I received my Master of Science degree from the University of Kansas in 1961. I had probably sent out more resumes by that time than any other student in history. With about 100 credits in geologic

1

science courses, I was sure that something would eventually turn up; ultimately it did, and it started a more fascinating geologic journey for me than I ever dared to dream! The fact that I would spend more than half of my geologic career working in the Jackson Hole region in northwest Wyoming and in the shadow of the Grand Teton Range was far beyond anything I could have ever imagined!

The summer jobs that I had during my student days were very instrumental in helping me in my search for employment. During the summers of 1957 and 1958, I worked for the Wisconsin Geological Survey and Wisconsin Highway Commission. The work involved digging test pits for sand and gravel deposits along the projected routes of the new interstate highways that were being built in Wisconsin. The main concern of the Highway Commission was that contractors would make fraudulent bids claiming that there were no sand and gravel deposits within 20 to 30 miles of the road project. Consequently, the contractors would need to haul materials and the construction costs of the final road projects would be much higher.

My working on these road projects had been a dream summer job for me in 1957 and 1958. The pay was $300 per month plus expenses. I had believed in the spring of 1959 that it was a foregone conclusion that I would be offered a job as chief of one of the crews since I had been actively involved the past two years. However, I was no longer a student at the University of Wisconsin in 1959 and this apparently made a difference with the professor who ran the program for the state. When the letter from the Highway Commission arrived, it hit me like a stone wall. It said very simply, "We are unable to offer you a summer position this year."

I was now in crisis mode since I was in great need of a summer job. I went to the bulletin board in the geology department at the University of Illinois and scanned it looking for any kind of a job. There was only one job announcement, that for a Museum Aide at the Smithsonian in Washington, D.C. The job would be in the paleontological section of the Natural History Museum. I went to the post office immediately on that Friday and filled out an application and sent it airmail. A week later I received a letter saying, "You have been selected..."

As you might expect, the job of a Museum Aide cannot be exciting, even if it is in the Natural History Museum. My job was to help clarify the nomenclature of some of the invertebrate fossil groups that inhabited the museum, a rather mundane process. My immediate supervisor had a Ph.D. from Cambridge University in England and was a strong atheist. The turtle man who was there part-time (he studied fossil turtles) was a strong Christian advocate and their arguments at lunchtime were quite heated. The biggest excitement in the museum occurred one night when the guards on the first floor held a quick draw contest and shot a number of bullet holes in the African elephant that was a prominent display in the lobby.

A few times after work I got off the elevator on the lower floors and was always amused by what I saw. The supervisors, who were white, had work areas that were quite nice. As soon as they left in the afternoon, the janitors, who were black (the term African American would not come into usage until much later) would occupy these same desks with their feet prominently displayed on top of the desks;

they now owned the building. It was always a slight shock when I stepped out of the building because the sidewalks, streets, and buses were crowded with women, women, women everywhere. The majority of the workers in the government were secretaries and they flooded the area at quitting time.

It was difficult to find housing on a temporary basis in D.C. because of the influx of foreign nationals who worked in the diplomatic area. At first I tried the YMCA where I had a single room; then I switched to a double room which was much cheaper. I was taking a taxi to an apartment complex trying to find a more convenient room when the taxi driver drove parallel for several blocks with a very nice Cadillac convertible. A slender guy with bushy hair and wearing a light blue suit was driving it. The taxi driver turned to me at a stoplight and said, "That guy is going to be president of the United States someday. He's Senator John F. Kennedy from Massachusetts." It was only later that I realized that I had gotten a glimpse of Camelot!

My first and only roommate at the YMCA turned out to be quite an interesting figure. Neil Buck was a former navy pilot who crash landed his fighter plane while on a training mission in Panama because his radio man didn't bring along his parachute. He managed to haul the guy out of the burning plane and in the process suffered severe injuries that kept him hospitalized over 17 months. In civilian life he joined the Richmond police force. In that job he determined that some of his superior officers were into the drug scene. After he testified in court against them a contract was put out on his life. He then moved to D.C. where he worked part-time for a detective agency. When I first met him, he was in the process of running a business that was engaged in painting various tract housing developments.

He was at loose ends when I met him because his painting crew had gone to New York City and had not returned. Later in the summer he found out that they were all in the hospital with severe injuries because they had been involved in a major traffic accident. After looking for a room for several days, I ran into Neil Buck on a street near the YMCA and he wanted to know if I was interested in rooming together at the Pinewood Hotel near DuPont Circle, a fairly nice area. Since it meant free rent, I immediately said yes. It turned out he had some connections at that hotel and became manager.

In 1959 the Pinewood Hotel was not a top of the line hotel. Neil Buck, now manager, constantly had problems with winos wandering into the hotel. They drank cheap booze, like vanilla extract that had a 30% alcohol content, which made the hotel frequently smell like a bakery. In fact, it got so bad you merely had to step into one of the hallways and you knew instantly that one of the winos had been there. Prostitution was always a big problem. There was one in particular that would rent a room and then spend the night pounding on hotel doors waking up the tenants as she looked for business.

A serious incident in the hotel occurred several weeks before I was going to return to graduate school. A group of four to five men came into the hotel lobby at night and were drinking and making a lot of noise. Neil Buck told them to quiet down and called the police, and they came and escorted the men out of the hotel. Several hours later we heard car doors slamming and the same group coming towards the front entrance. I went up and locked the front glass door. The group proceeded to yell and scream and pound their fists on the front door, but they couldn't get in the hotel. Several days later the men turned up again and the main antagonist wanted a fist fight to avenge his honor. Neil Buck took off his glasses, gave them to me, and the fight was on. It ended up with Neil Buck having the kid in a neck hold and threatening to break his neck. The next day the kid came back and apologized and wanted no charges filed because he was in a training program for the U.S. Foreign Service and did not want the incident known.

In the spring of 1961, I returned to Washington, D.C., looking for a job and drove past the Pinewood Hotel. It had been newly repainted and had a very distinguished group of foreign guests standing in front of it. I stayed in another hotel for several days. When I had time, I placed a call to Neil Buck in Richmond and he promptly drove up to have a late afternoon coffee break with me. As I left D.C., I thought that this was probably the last time I would see him. Surprisingly, several years later, I received a letter from him that indicated he had a number of personal problems.

The letter was about four pages long and it was obvious he had been drinking when he wrote it. Later that year I sent a letter to the detective agency requesting his address. I received a shocking reply; Neil Buck had committed suicide! He had taken his truck out in the hills around Richmond and had never come back. Nobody had found him or his truck for six months and animals had scattered his remains over a fairly large area. As I tried to remember his letter, he was very disturbed that his second wife had become a drug addict; in addition he had developed bone cancer in the hip fractures he suffered in the plane accident in 1944.

In retrospect, the museum job in Washington, D.C. was not a very exciting entry into the geologic profession. However, it was largely instrumental in obtaining my next summer job in Alaska as the supervisor in the museum had given me a great recommendation!

Chapter 2

ALASKAN DAYS

A Russian Invasion Must be Stopped

As the airplane delay stretched into several hours in Madison, Wisconsin, I had time to reflect on my job for the summer. It was early June of 1960 and I was headed for Alaska. My main concern had been getting enough money for the field season, but my father had lent me three hundred dollars until I got my first paycheck. My fellow classmates in geology at the University of Kansas had also tried to enlighten me about the Alaskan wilderness by telling me numerous stories. For some reason, there was one story that seemed to stick in my memory:

> **The young greenhorn had traveled to Alaska and built his log cabin on a lake in a very remote area far away from civilization. But it was very lonely on that lake shore, so far, far away from anything. One day, an old prospector turned up and invited him to a party. The young greenhorn was so ecstatic that he kept saying over and over, "Yes, yes, I would love to come."**

The old prospector scratched his wrinkled skin through a hole in his torn blue jeans and said, "Are you sure you want to come? There will probably be a lot of drinking and cussing." The greenhorn said, "No problem." The old prospector scratched his beard as he sat on a log and said, "There is going to be a lot of fighting going on." The greenhorn said, "No problem."

The old prospector thought some more. Finally he said, "There is going to be a lot of loving too, maybe too much for you." With that comment, the greenhorn almost exploded. "Look," he said, "I have been up here for six months all by myself. How many girls are going to be there?"

The old prospector scratched his white beard one more time as he pondered the question while spinning his false teeth on a stick. Finally he said, "Well, I think it will probably be only you and me."

Clearance for takeoff finally came and the turboprop bounced down the runway, heading for Seattle, Washington. It was going to be a long flight, stopping at three or four airports along the way and wouldn't reach Seattle until the next morning. As the plane made a big turn in the early evening over Madison, I looked down at the lights of the city. It was the first time I had ever been on a plane. It was hard to believe that for the next several months I was going to see a lot of Alaska from a small plane!

The plane flight to Seattle, where I was going to board a Boeing 707 for Alaska, also proved to be a much needed solace for my soul. I hadn't completed my work for my

Master's Degree at Kansas that spring like I had hoped. When I initially took some entrance exam tests to determine what course work I needed for my degree, I had committed a very elementary mistake on one of the tests. I had drawn several topographic 20 foot contour lines through a lake — not a good thing if you wanted the lake to hold water! Therefore, I earned the enmity of this one professor, especially when I didn't take his courses.

Things came to a head in the oral examination given to Master's Degree candidates. I had spent 30 hours without sleep getting my fossil plates ready for my thesis. I was tired and very exhausted. I had known that several of the students of this professor in the past year had spent maybe 10 minutes or less in an oral "examination" and basically were given a free ride. In a heated exchange with this professor, he made a very condescending remark. "Where did you get your degree from?" It was meant as an insult, the insinuation being, "How and why did anybody give you a degree in geology?" Looking across the table at him, I learned forward a little and replied, "Wisconsin. Where did you get yours?" Things came to a head very quickly after this exchange. Frankly, I didn't really care: I was going to Alaska!

The plane landed around 8:00 a.m. in Seattle, and I was the first one escorted off the plane onto the runway via a moveable ramp. Another uniformed male flight attendant immediately grabbed me and pointed to a Boeing 707 parked nearby and said, "Go and get on that plane." When I made myself comfortable in a front row seat, somebody tapped me on the shoulder and it was my boss for the summer, Donald Nichols of the U.S. Geological Survey. His comment was simply, "Welcome to Alaska." Apparently the Boeing jet had been delayed an hour so he had requested that any available seat be found for me on his plane if the circumstances permitted it.

Upon landing in Anchorage, I found that my one and only suitcase hadn't made the transfer to the Boeing jet. We did find our "horse" for the summer, a 1950 Jeep that another geologist had placed at the airport for our use. It was not in good shape, apparently it had been taken out of a sand bank in Arizona and shipped to Alaska for our summer project. Upon stopping in Palmer, I learned firsthand about the expense of living in Alaska. In the lower 48, a hamburger cost 29 cents. The first one I ate in Alaska cost $2.50.

Our destination was Meier's Lodge at milepost 170 on the Richardson Highway. At this time in 1960, the Richardson was the only road connection between Anchorage and Fairbanks. A much needed more direct highway between these two cities was completed in the early 1970s. Meier's Lodge, however, was at the exact location we needed for our work.

Our four-hour driving time to Meier's Lodge gave me the needed time to catch up on the summer project. The American military was very much concerned about a Russian invasion over the northern polar region. Big military exercises were to be conducted the next winter in 1961–62. The need here was to make terrain or slope maps of the area where the exercises were to be held. These maps were to be prepared based on the slope angles of various glacial features. Superimposed on these colored maps were different patterns showing the different types of vegetation.

The whole concept came down to a color code. Green areas with a certain pattern meant an 18-year-old soldier driving his tank was safe, whereas yellow indicated that he had to proceed with caution. A red color in an area meant, "Sorry, you just died as your tank tipped over several times." Although this all seems elementary, winter snow could hide a lot of dangerous features easily seen if the snow was not there.

As Nichols described the project, I thought that he seemed very qualified to lead this type of project. My feelings were well founded. During the Cuban Missile crisis about a year later, he was one of five engineering geologists of the U.S. Geological Survey who went behind the "Green Door" for the American military. Plans were being made in secret to send American Troops into Cuba if the crisis continued. These five geologists were using aerial photographs to locate landing areas for troops and heavy equipment. Fortunately, the crisis ended before this became necessary.

Arriving at Meier's Lodge, I found it to be an old rickety log building with two separate log cabins for rent. We rented the larger of the two cabins since it had two rooms in it with the larger having additional space where Nichols could work on his maps. When the proprietor learned of our project, he was quite amused. Some military field exercises had started the year before. He said, "Those young 18-year-old kids driving those tanks came in here in the middle of the winter frozen stiff as a board and had turned completely blue. Those tanks didn't have any heaters in them."

I was worried that I would never get my missing suitcase, but finally another geologist turned up with it. It took a week to get settled in. For me, money was the main problem. The cabin was costing $200 per month. We split the cost, but Nichols was getting per diem and I was not. With only $300 per month, that did not leave me much for food or anything else. Fortunately, we could go down to Anchorage and buy food supplies at the PX on the Air Force base since we were on a military project. To this day, I cannot stand the taste of tuna from a tin can as that was what we took out into the field to eat.

A small lake with a dock was across the road from the lodge where a small float plane could land. Nichols eventually located a bush pilot who agreed to fly us back and forth from the lodge to our project areas.

The view from the edge of a lake in the interior of the Alaskan wilderness as our float plane is leaving us. We will spend the next four to five days by ourselves without any method of communication with the outside world.

Before we actually started field work, we needed to collect field gear. We took the Jeep into Fairbanks and obtained equipment from the Alaskan Branch of the U.S. Geological Survey. At that time, Fairbanks was a rather small city, still having some gravel roads. Later that summer, we took another trip into Fairbanks, and I got a chance to walk around the town. In several areas I saw a tall, sunburned man in a khaki shirt and pants walking on the sidewalk with a very heavyset Eskimo woman walking about ten feet behind him with her five children. At first I didn't believe she was with him, but when I saw this scene repeated several times, it was apparent that they were together. I could only guess that he was a gold prospector who had "gone native."

It was on that same mid-summer trip that I wandered into a surplus shop in Fairbanks and saw several pieces of ivory tusks in a display case that had come from a mastodon. I was told that the tusks were found 30 feet below a river bed during gold dredging operations and were estimated to be about 18,000 to 20,000 years old. I bought the smaller of the two pieces for $25 and now use it as a paper weight even though it weighs about twelve pounds.

The ivory tusk is rather unique in that it is in almost perfect shape considering where it was found. It is about twelve inches in length, and six to seven inches in diameter. One end is completely solid with a conical cavity extending to the other end, leaving only a one-inch rim. The cavity is believed to be where the muscle attachment held the tusk to the mastodon. Crenulations run the length of the slightly curved piece. The dentine portion of the tusk was apparently stripped off to expose the very beautiful white

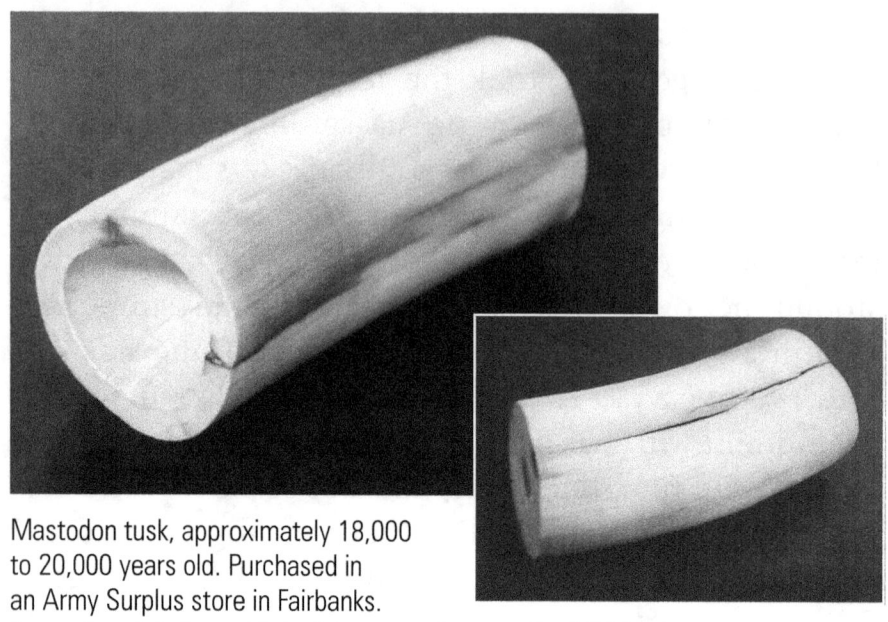

Mastodon tusk, approximately 18,000
to 20,000 years old. Purchased in
an Army Surplus store in Fairbanks.
It was found during gold-dredging operations south of Fairbanks.

ivory. Both the top and bottom of the piece are finely cut and beautifully polished indicating that somebody had spent considerable time and effort working on it. It is truly a museum piece.

In order to understand our project, one must have some understanding about glacial geology. This area in central Alaska south of the Alaskan Range had been covered in the past by a tremendously thick ice sheet at least several thousand feet thick. Streams that flowed southward underneath the ice sheet did not deposit their sediment load downward into the permafrost. The erosion effect was upward into the ice, which resulted in the deposition of material in that direction. When the ice sheet finally melted, the former stream deposits, called eskers, now existed as long ridges that in some cases extend for miles. These long ridges can

16

reach up to 50 feet in height and 100 feet in width. Hence, when we finally started out field work, we would try to find an "esker" going our way. The top was usually bare due to wind erosion and little soil, and mosquitoes were usually at a minimum. Nichols thought that it was the 3,000 foot elevation in the area that kept the mosquitoes from being as much of a problem as they were on the coast.

Eskers are a buildup of river sediment underneath a massive ice sheet. When the ice sheet melted, the collapse of the ice walls allowed the sediment to assume its current shape.

Of equal interest were the huge terraces that resulted when the ice sheets stopped their advance and started to melt. Melting of the ice sheet for a long time at the same rate the ice was moving resulted in stagnation of the ice front, with the material being dumped in one locality. An ice sheet does not push material ahead of it like a bulldozer, but instead carries material within the ice. These glacial features and resulting sharp slopes, plus permafrost and small lakes that filled numerous small depressions as well as different types of vegetation, would pose many problems for moving tanks and other heavy equipment during military maneuvers.

When I had worked on summer projects for the Wisconsin Geological Survey, eskers also existed in our project areas. However, they had been mined out at one time for sand and gravel and were now water filled ditches several miles in length. Of more interest to a geology student was the glacial grooving and polishing of limestone beds around some of the limestone quarries in Wisconsin. Quarry operators had cleared off soil next to existing quarries in preparation to expand quarry operations. The glacial features on the limestone surfaces were always polished and grooved in the direction of ice movement. If the surfaces had been flat you could have danced on them, but you might have slipped and fallen down!

The field season in Alaska started in earnest upon our return to Meier's Lodge from Fairbanks. The bush pilot that we used flew a Super Cub plane which didn't have very much interior space. Our two backpacks were stuffed into the tail of the plane. I was next, and Nichols sat on my legs in back of the pilot. I could barely see out of a small window when I leaned forward. I often wondered what would happen in a plane crash. It was rather disconcerting on many trips to look out the small window and see small gravel ridges apparently only a few feet away that seemed very level with the plane window.

Landings on a lake were always interesting. The pilot would circle the small lake several times to make sure there were no logs or dead animals in the water where he wanted to land. On the approach run he would throttle down the engine; at first there was no sound as the plane seemed to be suspended in the air. As he got closer to landing, one could hear the low purr of the engine. After landing, we unloaded our backpacks from the plane and bid our pilot a fond goodbye as we would not see him again for another 4 to 5 days. We were now on our own in the Alaskan wilderness. We had no radios, no way of communicating with the outside world in case one of us got hurt. This was the way it was in the year 1960.

We had arranged for our field trips to be made in a somewhat semi-circular path. The pilot would drop us off on one lake and we would have him pick us up on another lake. We each carried a big backpack with our food, sleeping bags, and other gear. Our tent only weighed about five pounds and measured about 7'x7' when put up. As the field assistant, I always got to carry the tent! All of our food was carried in tin cans, thus making it very easy to prepare; we believed this was necessary to prevent bear attacks.

Of special interest was the rain gear we had picked up in Fairbanks. Both the jacket and pants were made of a very tightly woven poplin fabric that would shed water. Locally this rain gear was called "tin clothes." They were much better than plastic rain gear that would tend to hold in moisture. We carried the rain gear in our packs and only pulled them out when we saw rain clouds in the distance. There was a lot of cloudy weather but we never had any extensive rainfall.

In Bear Country, one always had to keep a sharp lookout!

The only protection we had against bear attacks was a Smith & Wesson .44 Magnum revolver that Nichols had brought from Washington. It weighed five pounds with ammunition, and since I was the field assistant I got to carry that gun! Space limitations on the float plane had mandated something smaller than a rifle. It was an amazingly accurate gun. One time we had to wait almost five hours for the pilot to pick us up because he set down on the wrong lake. Nichols decided to shoot at some ducks on the lake and came close to hitting them at 100 yards or more, missing them by a foot several times.

The Smith & Wesson revolver was a very powerful gun. Sometime during the middle of the summer a story reached us at the lodge about some tourist trying a "quick draw" with the same type of gun. The front sight of the gun caught on the top of the holster and the gun went off upwards, hitting the individual's shoulder and basically severing it from his body. After this report I felt more confident if we had to protect ourselves against a bear attack.

Surprisingly, we never did see any bears. Only once did we find a sign of a grizzly. We were hiking across a low area containing a lot of low scrub trees when we came upon a thicket of much larger vegetation and found a big opening underneath the canopy. Huge bear foot prints almost as big as dinner plates were scattered across the ground. One side of the clearing also showed where the bear had been making his bed. Everything looked very fresh!

Surveying these bear prints, I told Nichols, "If we see him, I'm giving you the gun and I'm going to make a run for it. I don't care how big he is and how fast he is, I can outrun him." Fortunately that didn't come to pass.

We tried to keep a normal work schedule in our field work despite the ever-present daylight. One night in camp, Nichols had his watch stop, so we tried to guess the correct time. Getting up we set his watch to the suspected time and did our field work before ending up at the lake where the pilot was to pick us up. When the pilot finally arrived, we learned the correct time and determined that we had gotten up at 3:00 a.m. that morning.

Odd things occasionally occurred. One night we had set up our tent next to a small stream. While trying to sleep, I kept being wakened up by noise that sounded like a bunch of ducks in the nearby stream. Being quite tired, I did not want to get up and chase them away. Making camp in about the same place the next night, the noise started all over again. This time, I got up, put on my boots and clothes and prepared to chase the ducks away. Exiting the tent, I stared in amazement. There were no ducks or any kind of birds. The whole small stream was literally boiling with small fish jumping for mosquitoes!

We drank water directly from the streams with no purification of any kind. Nichols' philosophy was that there was nothing to pollute the water. He must have been right because I had no GI problems the entire time we were in the field. The biggest obstacle we had in the field came when we needed to cross streams. We had taken the temperature of water coming directly out of the snow fields of the Alaskan Range and found the temperature to be about 34 degrees. Consequently, we usually waded the small streams keeping our field boots on.

There was one stream we reached that was about 30 feet across. I decided to try something different. I took off my field boots, tied them together, and put them around my neck and proceeded to cross the stream wearing my big backpack. This was a big mistake as I was top heavy and it was almost impossible to keep my balance. In addition, the very slippery rock surfaces became dangerous and the uneven stream bottom hurt my feet. After struggling across, I later wondered what would have happened if I had lost my balance and fallen in the water and lost my field boots. Never again did I cross any streams by taking off my boots!

During the middle of the field season one of the other geologists working in the area hit a moose calf and severely damaged his car. Moose have long legs, no matter what their age. The car had slid underneath the moose calf, with the animal sliding up and over the front car hood. In the

<u>Not</u> the way to cross a stream! We usually waded a stream leaving our boots and clothes on. Because of the width of the stream, I took off my boots and hung them around my neck as I didn't know how deep the water was. If I had fallen due to the smooth rocks and uneven bottom and lost my boots, it could have been very serious.

process, the broken headlights, and probably the hood ornaments, had sliced the belly of the animal open. The moose hit the front windshield, completely destroying it. The inside of the car was completely covered with blood, guts, and what could be politely described as "manure." The geologist hadn't been seriously injured. He placed clear plastic sheeting over what had been the front windshield in order to drive the car. I got the joy of following him in our Jeep to where he could drop it off for repair. Seeing what the car looked like, I always wondered if the moose looked better than the car.

We did have some down time one week during the summer when the pilot had some mechanical problems with his plane. Nichols decided that gave him a little time for an update on his "mud volcanoes." These features were located on the other side of the Copper River, and with no road access, we needed to take a rubber raft across the river. Another geologist and his assistant joined us for this venture as the river was wide and deep and at this time of the year was running very fast. We put our Jeep a quarter of a mile up river on this side of the river from where we wanted to land on the other side. The other Jeep was placed a quarter of a mile down river on this side from where we expected to end up on our return.

The rubber raft was probably 10 feet long and looked sturdy enough and, of course, we were all wearing life jackets — just in case, I untied my field boots. The ride started off slowly as we hit some sand bars, but then we hit the main river. The two geologists were paddling like mad as the raft went up and down three to four feet because of the rapids. The other assistant and I had no paddles and we were hanging on for dear life. Several times we were in the bottom of a trough and water swells were so high we couldn't see the other shore. The next second we seemed to be sitting on top of the world. It was a roller coaster ride to end all rides. Finally, after what seemed like an eternity, we reached the other shore. It turned out we actually needed a quarter-mile allowance to the other side where we wanted to be. When I thought of the return trip, I wondered why I had agreed to come along.

The term "mud volcanoes" was probably not the best term to adequately describe these geologic features. They were really huge extended mounds formed by the upward movement of water and different volcanic gases which were bubbling to the surface carrying some sediment with them.

Two different types of "mud volcanoes."

No vegetation was on the mounds, but with the extensive moisture we would sink several inches into the ground. Nichols had written several articles about these "mud volcanoes" and wanted to collect additional gas samples to see if the composition of the gases had changed. Frankly, I couldn't get too excited about them. I was looking forward to the return ride across the river. Again, it took lots of paddling to get back to the original side, and it took exactly a quarter-mile to hit the Jeep location.

Native fish trap. Big paddle wheels are typical of these types of traps.

Coming back towards Meier's Lodge, we stopped at a small settlement along the river. As I wandered among some of the old log cabins so typical of Alaska, I went over to the ridge overlooking the river. Down below was a big native fish trap slowly turning in the water. It was typical of most of these types of traps with big paddle wheels turning because of the current. A small boy about 8–10 years of age came up from the river carrying a large salmon. Holding it as high as he could, the tail was still dragging on the ground. Curious, I asked him if he wanted to sell his fish. Yes, he would sell his fish for 50 cents. Thinking it over I decided to let him keep his big fish.

In late August the end of the field season was coming up fast. Nichols wanted to fly over the area in a much bigger two-engine plane for an overall view. I decided to decline the invitation to ride along. I said, "I won't ride in any plane where you have to wear a parachute." The military also furnished a big helicopter for several days. Again, I turned down the invitation to ride along. When I looked at the 30 foot long rotors slowly turning as the engine was cranking away for 15–20 minutes in the morning hours, I decided that float planes were really more my style.

On the last night at Meier's Lodge, I walked up and down the Richardson Highway in front of the lodge for several hours. The air was fresh and still and the night had started getting a little darker. This night in particular the stars were seemingly so close one could almost touch them. In the far distance the silence was occasionally broken by a loon giving out its lonely cry over the lake. I took one very long, deep breath before going back into the lodge; I knew for sure that my soul had been fully restored!

The return flight to Seattle went off with no problems. With hardly any money left, I was forced to take the train back to the Midwest. In Minneapolis I had to change trains to Madison. While I sacked out in the train station at night, the station master started yelling at me as he ran down the length of the depot. My appearance had made him believe that I was a hobo in for the night. My old worn out pants held up by suspenders, my faded torn shirt, and my full length red beard could have confused him. As I handed him my ticket, he just stood there for what seemed like several minutes, obviously dumbstruck!

In June of 2008 I had to return to Alaska to see what I had missed in 1960. I put my 23-foot motor home on the ferry at Bellingham, Washington, for a ride up the Inland Passage. The ride to Skagway cost 2,800 dollars. I made a full circle of Alaska before coming back up the Richardson Highway. The highway was now in sad need of repair. Meier's Lodge had burned down several years after I had left Alaska, but two cabins that were still there were apparently the same ones that had been there in 1960. The lake was completely hidden from view by extensive vegetation and the dock was long gone. Rain was starting to come down as I walked a short distance up and down the highway near where the lodge had once stood. My memories were really the only things that remained!

Chapter 3

THE ODYSSEY BEGINS

In 1958, I had been shocked by the lack of job opportunities after graduation from college. I was in even greater shock when I completed my course work for my Master's Degree at the University of Kansas and the job market was still terrible! I was now 24 years old and had spent the equivalent of almost three years in graduate school. Before I left Kansas in January of 1961, I tried enrolling in the Education Department with the idea of becoming a school teacher. I gave that idea up very quickly!

When I returned to Wisconsin in January of 1961 to stay temporarily with my father, I decided to try another profession, that of an automobile insurance adjuster. I spent three days riding around with an insurance adjuster visiting what seemed like every junk yard and body shop within 50 miles of Madison. The insurance company was not interested in repairing damaged cars with new parts. They indicated that used parts would be used whenever possible as using new parts would cost too much. After the three-day period,

I had a meeting with the office manager and he declined to offer me a job. He said, "After much consideration, we won't offer you a job. If a job comes along in your specialty, you will probably quit your job with this company after we've spent considerable time and effort training you." I had to admit he was right — I would have quit!

Consequently, I tried a number of temporary jobs. I raked leaves for three weeks at Indianola, a youth camp east of Lake Mendota for 75 cents per hour. When that job came to an end, I found a job unloading a semi truck of 100 pound fertilizer sacks for $1 per hour. The truck was backed up to an unloading dock and another student and I spend the day unloading it by hand, with no wheeled hand truck provided to us. By the end of the first day, I was so numb I couldn't decide if I had usable body parts. I went home and soaked in the bathtub for almost an hour. It didn't help — I could barely move for three days.

I found other odd jobs in the Madison area. I helped to unload a train box car of finished lumber, at 80 cents per hour. I worked for a moving company for several weeks moving trunks from sorority houses for students moving back home after the semester ended. The call would go out, "Man on Floor," and there would be a rush of scantily clad females heading for cover — sometimes they didn't hear the call! The best temporary job that I had for three weeks was bolting together small chicken brooder houses from kits that the University of Wisconsin Extension Service was going to use for experimental projects. While this job paid $1 per hour, it was obvious that I had to find a permanent job.

In 1959, I had taken a written exam given by the

U.S. Geological Survey for placement on a Register of Job Applicants. I had passed that test which had partly led to my job in Alaska in 1960. The equivalent of that test today would be the SAT given to high school students who are college bound. The test given by the Survey was concerned only with geology and consisted of 100 multiple choice questions, with 30 of the questions related to a geologic map exercise.

The map exercise was rather unique. All the map symbols showing the identity of the rock units were removed, as well as all the strike and dips of the rock units. In order to answer the questions, one had to figure out the geologic structure in the map area. I came to the conclusion that the structure was a syncline on a high plateau. When I came out of the room where the test was held, the other students were proclaiming that it was something else. I passed the test; they did not.

I concluded in the spring of 1961 that my best chance for a permanent job consisted in getting a job with the U.S. Geological Survey because of the downturn in the oil and gas industry. I packed my bags in my 1952 Chevy and headed to Washington, D.C., to try and arrange a job interview. After much pounding on doors, I finally managed to land an interview with the Chief of the Geologic Division. He indicated that very few geologists were being hired by that Division. He said, "We are going to hire only five Ph.D. geologists this year to work on our Kentucky mapping project. The Conservation Division is going to hire a small number of geologists. Try them."

The interview with that group was not very positive. The Branch Chief said, "It all depends where you are on the

Register." He called the Personnel Office and determined I was in the middle of the list. He said, "We can only offer you a job if some of the candidates ahead of you will turn down our offer of employment. As we will drop them off our list, we may be able to offer you a job."

Finally, an offer of employment with the Conservation Division of the U.S. Geological Survey came in the middle of May for $6,345 per year working out of the Denver, Colorado, office at the Denver Federal Center. Of course, I immediately accepted! It was fortunate that the offer came at the right time. The day before I had received an offer in the mail for a job with the Corps of Engineers in Missouri to work on building dams for flood control. It was also a good opportunity, but the one in Colorado was much better — it was the one I really wanted.

. . . .

As I sat in the office of the Regional Geologist with ten other young geologists in early July of 1961, I learned my field area had been changed. Central Colorado had been designated originally, but now it had been changed to southeast Idaho, apparently one mountain range removed from the Jackson Hole area of northwest Wyoming. After I listened to the discussion about the mapping program, I realized I really owed my job to former president Theodore "Teddy" Roosevelt. During his presidency from 1901 to 1909, he had withdrawn thousands of acres from the public domain to prevent private interests from exploiting the federal lands for their valuable mineral deposits. The geologic mapping program was designed to determine if valuable mineral occurrences existed on the withdrawn lands and then classify those lands as to their mineral values. Roosevelt's interest in protecting the natural resources of the United States was genuine, and I felt honored to be included in any program that would accomplish that goal.

As I drove across southern Wyoming on the way to my field area in southeast Idaho, I had time to reflect on how history determines future events. In particular, I was thinking how the life and times of Theodore Roosevelt was going to determine part of my career. Although he was gifted in many areas, it was his famed charge up San Juan Hill in the Spanish-American War that helped propel him into the presidency. That brave military charge was based partly on a misconception by the American public.

The American forces on San Juan Hill had been hopelessly pinned down by the Spanish troops. The Spanish troops were well dug in and had the best repeating rifles available and also used smokeless powder which did not reveal their position. When every American soldier or their Cuban allies fired their black powder rifles, a cloud of black smoke would instantly reveal their position and would draw immense return rifle or cannon fire. It was the arrival of four multi-barrel Gatling guns that made the difference for the American side. Each of these guns could fire 600 rounds of ammunition per minute and with their combined fire power could place over 2000 rounds every minute on the Spanish fortifications. This immense fire power made the final American assault succeed.

Roosevelt's description of the final battle indicates he came very close to becoming killed or seriously wounded many times. Soldiers were being killed or badly injured all around him, but Roosevelt only suffered several minor wounds. If he had been killed, his presidency would never have happened. The National Park system would have been much delayed, and the thousands of acres he withdrew from the public domain for economic evaluation would never have happened. The geologic mapping that I was going to be involved in would never have existed! As I made the turn at Rock Springs, Wyoming, and headed north towards my field area, I wondered how many of my relatives would ever believe that I owed part of my career to "Teddy" Roosevelt and four Gatling guns!

. . . .

Sometimes a person is involved in history without realizing that he is experiencing history! When I saw John F. Kennedy in 1959 in Washington, D.C., little did I realize I saw part of Camelot. When I joined the U.S. Geological Survey, little did I realize the type of science group I had joined. The Geological Survey by the early 1960s had gained the reputation as the premier earth science organization in the world. The agency had tried to hire the best earth scientists it could find for its research projects. I was fortunate to be able to work within that type of organization.

The door was always open to you inside the Survey, so it was possible to discuss any aspect of earth science with highly regarded scientists. You might be walking down a hallway and see a highly respected scientist from a university or from Europe. I could sit down at a lunch table in the cafeteria and listen to a very spiritual argument about continental drift, hot blooded dinosaurs, or when the next major eruption might occur in Yellowstone. It was an atmosphere very conducive to scientific learning. I was fortunate because this kind of atmosphere does not exist in an oil company.

The U.S. Geological Survey in 1961 consisted of four divisions — Geologic, Conservation, Water, and Topography. The Conservation Division got involved in the mapping program because it had evolved to the point where it managed the leasing and evaluation of federal leasable minerals. The Geologic Division prided itself in being more involved in the scientific aspects of geology. The mapping program had actually started in 1960 when older geologists in the Geologic Division were transferred to the Conservation Division. The program expansion involved hiring another 13 young geologists the following year.

There were about 30 field geologists at any one time in the Denver office to handle this monumental task facing the Division. The entire office staff was based in Building 25 at the Federal Center in what became part of the City of Lakewood. The states that had most of the withdrawn lands were Colorado, Idaho, Nevada, Utah, and Wyoming. Most of the field activity was based in Wyoming and Colorado, with about 20 geologists who had some field projects in Wyoming at one time or another. Colorado had about twelve geologists who had some project work in that state. At the end of the mapping program in the late 1970s, I was the only geologist actively working in Wyoming. Many geologists had retired or were deceased; others were told to get "in the line of progression" and move to office positions. I turned down chances to move up the ladder as the Jackson Hole and Grand Teton areas had gotten into my blood. I wanted to stay as long as I could. I had one advantage over everybody else — I was the youngest geologist in the organization.

The mapping effort was simple enough. Geologists would map the geology on aerial photographs during the field season between May and October of each year. The data would usually be compiled on a 7½-minute topographic map during the offseason or winter months. Each map averaged about 54 square miles in area. After compilation, the geologic maps would go through two technical reviews and then the map editor. After this effort, the maps were sent to the Washington headquarters for publication. A copy would always be sent to you for your review so you could sign off on the final printing. Depending on the complexity of the project, the average time from the start of mapping to map publication was about two years. Arguments could crop up between geologists about some aspect of geologic interpretation that might cause an additional delay.

The amount of work turned out was astronomical with over 8,000 square miles covered by our geologic maps. I ended up with my name on geologic maps that covered around 1,300 square miles of some of the most structurally complicated areas in North America. I used horses during the mapping of most of my field areas, which is why I put on about 6,000 miles on horseback. When I still manage to drive into the Jackson Hole areas, I always look at those high mountains and wonder how I managed to work in that rugged country.

. . . .

I will always remember my first field season. When I first arrived in the Swan Valley area in 1961, I started working with D.J. who was an avid horseman. The previous geologist working with D.J. had decided the hills and mountains of southeast Idaho were too much for a man his age, and he needed something more "gentle." He had been in World War II and was in the second wave to hit Omaha Beach and thereafter served as an advanced scout throughout the war effort. There was no road access except by horseback into the Snake River Range east of Swan Valley. That area needed to be worked in the next several years; he was a man who could see the handwriting on the wall.

My introduction into riding horses had proceeded fairly quickly. It was one thing to ride a horse on a wide, level trail on a dude ranch and quite something else to ride a horse off trail. After the first several days of riding, I started to place big sponges in the back pockets of my Levi's. This helped ease the transition into the saddle riding. The first mapping I was involved in was evaluating the phosphate deposits in two areas in the Caribou Range, which was west of the towns of Irwin and Swan Valley. This area had fairly low elevation and horses didn't need to be used every day.

My education about horses started early. I'm sure everybody in the Valley heard me start yelling at the horse when he bit me. Early in the field season I had been loading a small horse onto the bed of a Dodge 3/4 ton truck next to Snoopy, a five-year-old black mare I had been riding. Snoopy was considered to be one of the calmest, most gentle of horses. I had loaded her first. As I led the smaller horse onto the truck, Snoopy turned her head towards me with her eyes blazing red. I knew that the two horses had been kicking at each other in the corral.

The first time Snoopy turned her head, I stopped loading the other horse until she calmed down. At that point, I continued with the other horse and then it happened! Snoopy turned her head again and I saw the flash of white teeth and then felt the bite. Moving quickly, I got off the truck and closed the tailgate. I took off my heavy jacket and shirt and inspected the damage. It could have been worse as it was only a two-inch mark just above the nipple. I felt I knew why the Regional Geologist had given me a big grin as I was leaving Denver for Idaho when he said, "I understand the guy you are going to be working with uses horses in his field work."

When I later described the horse action to D.J., he replied, "Gee, I'm so surprised that the horse did that." I said, "You're not as surprised as I was!" It was almost 20 years before that bite mark went away.

The two areas or quadrangles we were mapping in the Caribou Range contained only phosphate resources, no coal deposits were present. An evaluation program had been set up the year before with trenches hand dug across the phosphate bearing intervals of the Permian Phosphoric Formation. This was the only rock unit that contained phosphate. Trenches were starting to be dug in all quadrangles that contained the phosphate bearing rock. Trenches were dug in the summer by college students with rock samples collected across the entire phosphate bearing interval. Samples were sent in for chemical analyses at the end of the field season. These analyses were eventually included on the geologic map when it was published.

Some of the ranchers I met in the area were great philosophers. I remember one old rancher who stopped to talk to me while I was banging on a rock outcrop late on a Friday afternoon. He wanted, obviously, to know what I was doing. After we had talked awhile he said, "You know, there's not much differences between us and those Russians; just don't pay your taxes one year and you will find out who owns your land, you or the government." I have often thought about that comment!

Petterson's store in Irwin, Idaho, in 1962. Gasoline is 33.9 cents per gallon.

I had lodging in an old slightly run-down building next to Petterson's Trading, a general store with groceries and gas pumps out in front. D.J. had persuaded Ernie Petterson who owned and ran the store to rent me the space as there was nothing else available. At the end of the day, being kind of stiff after riding horses all day, I would wander into the store to see what I could put on the evening menu. I would always stop and look at the T-Bone steaks in the display case and say, "Ernie, I don't see anything I like for dinner." Ernie would just grin and take me back into his big cooling unit in the back of the store that had a couple of beef halves hanging up. "What looks good to you," he would say. I would carefully survey all the possible choices and either point to a nice porterhouse or prime rib section and say, "I'll take one of those." Ernie would get his big meat saw down and cut me a nice 1½-inch or more thick steak. Considering that I had seldom seen steak during my college years, I would say, "Yeah, that one looks pretty good" — and it was!

Days would grow long in this rural environment. There was no TV service and the radio reception was poor. The Post Office was in the store, but mail was very infrequent. Late that summer, I still hadn't made it over the mountain range to Jackson, Wyoming. One weekend I was kidding one of the waitresses at Shangan's Lodge which had a small bar and grill, about going to Jackson with me some evening to see the lights of the city. She looked at me and immediately said, "Sure, I would love to go but first I have to ask my mother." She was a very attractive 20-year-old Mormon Student at Riggs College in Idaho.

Her mother did agree she could go to Jackson, so on a Saturday night we took off, not an easy drive over the mountain in my 1952 Chevy. From Swan Valley, one drives east over Pine Creek Pass to Victor, Idaho, and then over Teton Pass to Jackson. The Homer/Jeffro show at the Wort Hotel was a riot. The very next Monday I went over to see our horse packer near Driggs, Idaho, in Teton Valley. His very first comment was, "I heard you're dating one of the local Mormon girls." The distance between these two areas is over 50 miles and across one mountain range. Gossip sure seems to travel fast in these small Mormon valleys!

Every time I hear Johnny Cash and his wife June Carter sing, "I'm going to Jackson," I think of going on that very narrow winding road over Teton Pass in the dark with its many hairpin curves and steep grades that existed in 1961! The "Jackson" they were singing about of course, lies over 1,000 miles southeast of Jackson Hole, Wyoming!

Chapter 4

THE BLONDE FROM MUSCLE BEACH

I came back to Denver full of wisdom —
and not a dime in my pocket

Tragedy suddenly struck the Conservation Division in late fall of 1961. I had returned to Denver in September from my field area and apparently everything seemed to be fine. In December the Regional Geologist called an urgent meeting. An accounting error had been found and the Division would run out of funds to pay salaries. Arrangements had been made to transfer six of the geologists just hired to work in the Special Projects Branch of the Geologic Division that was working on the Nevada Test Site. I declined the opportunity to physically work on the Test Site as I was concerned about being around radioactive terrain. Consequently, I sat in an office on the Denver Federal Center examining well cuttings under a microscope and describing them in a Technical Letter.

Later that spring I was sent down to the Nevada Test Site for two weeks. After flying into Las Vegas, I was driven 60 miles northwest to Mercury, a rather dusty makeshift collection of buildings somewhat classified as a government town. I was given a badge that would indicate if I had been exposed to any radioactivity. The dormitory rooms were clean and small, but adequate. The small cafeteria served very reasonable food at a low cost. It was starting to get hot, and I longed for the green and rugged mountains around Jackson Hole and the Grand Tetons. The nights seemed to drag on forever in this hot, dry country.

After several days, a young guy came around the core lab where I was working and wanted to know if I would like to tour the Test Site. "Oh yes," I said, "I would love to tour the place." I jumped at the chance to get out of the building because of the heat. We climbed into an old Ford truck to start the inspection trip; a tour that I was going to regret. The paved road running throughout the Test Site was very, very narrow and there were big potholes and ruts all over the road surface. This resulted in the kid zooming over, around, and through all these pot holes while pointing out the various nuclear sites. The truck had no restraints and I was constantly thrown around inside the vehicle. What I could see when I got the chance to look seemed like a giant prairie dog town. Obviously, the government had tested a lot of nuclear bombs. There didn't seem to be any consistent pattern to where the nuclear devices had been set off — big pits and mounds were scattered all over the area.

Later on, several other geologists and I were given the chance to measure the diameter of the various rocks and the distance they were thrown from some of the blast hole locations. Fortunately, these sites were slightly radioactive and the project was scrapped. Sometime later the young guy turned up again and offered me another ride. I politely turned him down explaining that my timetable wouldn't permit it.

When I listened to several of the more senior geologists talk about the nuclear testing, I got somewhat upset. They discussed various nuclear tests that had vented to the atmosphere and how far the radioactive clouds had traveled. In many cases, the radioactive clouds had not only drifted across the entire United States, but also out over the Atlantic Ocean. Search planes stood at the ready to follow these radioactive clouds to determine how far over the Atlantic they extended. Different newspapers had also reported sheep dying in the Utah and Nevada areas. Reports were also surfacing about an apparent upswing of cancer occurrences in the United States. I vented my disgust at various times about why more of these nuclear explosions that had vented to the atmosphere were not publicly disclosed. I don't remember seeing any published account of planes tracking radioactive clouds across the United States in the 1940s and 1950s. Information like that was not readily disseminated.

. . . .

In early June of 1962, I stood in the office of the Chief Administrator for Special Projects in utter disbelief. It wasn't the fact that I had just been told I would have to drive 600 miles to Carlsbad, New Mexico, to do some work on an underground nuclear site called Project Gnome. It also wasn't the fact that it was 1,400 feet underground, but the fact that I would need to climb into a steel cage to be dropped to that depth to do nothing more than help clean off some tunnel walls! Although my assignment was supposed to be related to my job, I felt that it really was in retaliation for my disgust of the nuclear program. Since I was leaving at the end of my six month assignment, they were just going to find some menial type job for me just before I left.

When I arrived in Carlsbad, New Mexico, I found that the project location was a few miles south of that town. Headquarters had been established in Carlsbad and a van would take a group of individuals to the Test Site every morning. When I first arrived at the site location, I was surprised to see a very small building that housed the wire cage and the supporting structures. Off to one side of the building was a 10 foot steel drum with a one-inch steel cable wrapped around it, with the cable then leading to a pulley system that operated the cage. A diesel engine was attached to one side of the drum. On the engine sat the engineer, a young bearded bare-chested man with blue jeans tucked into his cowboy boots. Judging from his hairy exposure, I guessed he was somehow related to the missing link. I got into the mesh steel cage with several of the miners. One of them looked at me and said, "Ready to go to hell?" I nodded. The drop was 1,400 feet in about 60 seconds. Surprisingly, as the cage dropped, there was little noise and vibration. Every 200–300 feet there was a sudden burst of light indicating we had just passed another tunnel.

Arriving at the bottom of the shaft, I was surprised to see a fairly big lighted tunnel that would lead to the nuclear blast hole about a quarter of a mile away. It didn't seem that Project Gnome was a military project. It had been designed apparently to test how rock fractures produced by a nuclear explosion would help in the movement of oil and gas to a nearby well location.

Strain meters had been placed in the tunnel to measure the force generated by the nuclear explosion. The fracture patterns along the tunnel walls were to be correlated with the data received from the strain meters. The tunnel walls had to have the dust and other debris blown off in order to measure the orientation of the fracture patterns. After looking at the fracture patterns, my belief was that the data generated here would be of no value in other areas because of the rock types at these locations. The soft sediments at this location would have tended to muffle the blast effect and the data would not be applicable to areas with more consolidated rock types.

Our badges were checked every day for any exposure to radioactivity. All radioactivity was supposed to have ceased months before. Even so, several oil company geologists who had toured the facility several weeks before had their cameras confiscated because their film became radioactive.

The job was rapidly becoming ho-hum and exceedingly boring. So one day, with time on my hands, I decided to take a closer look at the blast hole itself. It sat at the very end of this tunnel complex with the entrance about 20 feet above the tunnel floor. After climbing up a rickety old wooden ladder, I could see a dome shaped cavity. Flood lights had been installed around the outer limit of the cavity and were pointed upward. The whole cavity was lit up and sparkled with salt crystals. The height of the cavity was probably 150 feet and about 100 feet wide. It had not been a big nuclear device.

I climbed into the cavity and stood at the very center. It was a magnificent sight. Suddenly reality struck. The temperature was hovering around 140 degrees. At first it had not been noticeable, but suddenly I could not get any air. I was slowly being choked. My glasses fogged over and finally I managed to drop to my knees and slowly crawl back to the ladder and fresh air. I was sitting in the tunnel shaking with cold when I asked a miner walking down the tunnel about the tunnel temperature. He looked at me kind of funny and said, "It's about 95 degrees."

Several days before I was to leave and return to the Conservation Division, the project supervisor called me over and said, "There is somebody here whom I would like you to meet." It was the engineer from Livermore Laboratories who had been involved in some of the engineering studies at the site. I stopped short. She was obviously very blonde, looking very nice in stark white coveralls with a white helmet not quite covering the cascade of blonde hair. Wearing earmuffs, she took them off to talk. It turned out she was quite unique in that she was apparently the first female who the miners had agreed to let work underground. Women underground were usually considered bad luck and were banned. However, I could see why they relented.

Over the next several days we had occasion to talk to each other, usually at lunch time. She lived near Muscle Beach, California. I was lamenting on the second day that Carlsbad was such a sleepy old western town and that there was nothing to do at night. The only activity I could find on the bulletin board at the hotel was a square dance tomorrow night. She brightened considerably. "I used to square dance," she said. "Oh," I said, "Really? Well, how about going dancing together?" I also said, "We might as well grab something to eat before the dance."

So we met for dinner at the small hotel where I was

staying. Although it was a small place, it apparently was known for its good food as it was highly recommended. The waiter suddenly appeared. "How about a drink before dinner?" he asked. The blonde really perked up and said, "Do you have these wines?" and proceeded to name off seven or eight different vintages along with the year they were bottled.

"Oh yes," the water replied. "We have some very nice French and Italian wines." I mentioned something about the prices. The water looked annoyed. "Well," he said, "The French is the best, starting at $40 per bottle. Also, the Italian wine is just as good and starts at just $30 per bottle. Of course with these wines, we sell them only by the bottle." Suddenly the blonde didn't look so interesting anymore!

I did some rapid mental calculations. A year ago I was making 75 cents to $1.00 per hour trying to exist on temporary jobs. French wine would have me working close to 50 hours just to pay for one bottle at that scale. Obviously, the Italian wine was a much better deal. After some thought, I mentioned that I really liked Italian, regarding it's more pleasant tasting and smoother on the pallet.

Dinner proceeded smoothly. I watched in fascination as the big bottle of wine slowly disappeared as she talked incessantly — following the brief moments as she carefully took small bites from her very rare filet mignon. As I sat there, I started to wonder, "Where in the hell is Muscle Beach anyway?" My hamburger was excellent!

The square dance was held in the gym of a small school a short distance away. Only three to four squares of dancers were present, but everybody was quite friendly, and I was having a good time. The blonde managed quite nicely, even if she knew only a few square dance calls. She would put up her hands, give a big smile to any man coming her way, and would continue to dance very well. I was thoroughly amazed!

However, she did become distraught as the evening wore on and commented that we should leave early. I agreed, and as she was getting in my car she was holding her head, saying she was starting to come down with one of her frequent migraines. She indicated she needed to lay down immediately because of the pain. I let her off at her motel. I wondered if she had given her phone number to the young guy in the black hat who she was constantly smiling at. No, that couldn't be it.

My trip back to Denver went smoothly enough. I woke up one of the other geologists at 4:00 a.m. to borrow some money — I had spent all I had the night before. I came back to Denver full of wisdom — and not a dime in my pocket. Only later did I learn the truth about her hasty departure. After our agreement to go out, she had run around asking all the older geologists if I was single and most importantly, was I a gentleman that could be trusted? Of course, they all lied!

Later that fall, I was walking down the hallway of Building 25 in the Federal Center when one of the other geologists hailed me down. "Hey Marvin," he called. "Project Gnome just made LOOK magazine; yes, it had. The picture was taken in an upward direction from the very bottom center of the cavity and was of poor quality because of the glare of the salt crystals. Of more interest was a small insert that showed a young blonde standing next to a counter at Livermore Laboratories. She had had her moment — she had made the big time.

I sighed. Little did I know that within the next several months I would be dating and then shortly engaged to one of the most stunning and attractive brunettes I had ever laid eyes on. Of course, you know the rest of the story.

Chapter 5

RETURN TO SWAN VALLEY

The flash of lightning and roll of thunder on a nearby mountain peak sounded so close that it had me ducking for cover. The horses thrashed around trying to break their halter ropes. I had been watching the approaching storm for several hours but it was moving much faster than I had anticipated. I looked at D.J. and yelled, "Let's get off this mountain top!" His answer was to just head for the horses as we couldn't waste any time. My return to Swan Valley was proving just as eventful as I had hoped. It didn't seem it would take me very long to forget the Nevada Test Site.

I learned that getting off a high mountain top in a hurry with a horse was actually fairly easy. You grab the end of the lead rope with your left hand and with your right hand grab the lead rope about 12 inches from the horse's head, and downhill you go in what could only be described as a "running walk." The slope is now wet and slippery but you have the halter rope to stabilize your "free fall." You are very aware that you have a 1,200 to 1,400 pound horse right behind you that could trample you to death. The horse is much more stable than you are because he has four legs and you don't. If you add in rain streaked eye glasses, loose rock, tall grass, talus rock piles, and steep slopes, and of course the rain, lightning, and thunder, you end up having a very interesting afternoon. On the way down my Brunton compass got ripped off my belt, and I had to write an explanation for the front office explaining how I possibly could have lost this piece of equipment.

Riding a horse can be a tricky proposition for any adult not having had that opportunity at a younger age. I grew up on a dairy farm in the southern part of Wisconsin and had used horses for farm work, but I had never ridden a horse until I started field work with D.J. I found out it could be dangerous work as I had to deal with very narrow and twisting trails, talus slopes, steep hillsides, slippery stream crossings, as well as downed timber in forested areas. In addition, I wore traction-type field boots which will tend to stick in a stirrup if the horse should slip or fall. If the weather moves in, it can compound the conditions under which I had to work. After being almost thrown from my saddle when I first started riding, I developed my own method of survival. While my left hand always held the reins, my right hand always rested on the saddle horn. This procedure prevented me from being thrown from a horse as I could always grab the saddle horn in a hurry. I was never thrown from a horse during my 6,000 miles on horseback.

A serious incident did happen in July of 1962 when D.J. was thrown from a horse. We were riding along Palisades Creek on a trail that was about 30 feet above the stream valley. The trail was winding its way between a rock outcrop on our left and low willows on our right which extended down to the stream bottom. Several boy scouts suddenly appeared along the trail in front of us with long fishing rods sticking above their trail. D.J.'s horse did a violent 180° turn with D.J. thrown down towards the stream bottom. I sat on my horse absolutely petrified. I had seen his feet come over the top of the willows near the stream bottom. He suddenly exploded from the willows as he climbed uphill yelling, "It's just my ear." Many years later I found out that he periodically complained about his bad back. He probably would have been killed instantly if the willows had not partly cushioned his fall.

Horses have a tendency to "puff" themselves up when they are being saddled, making it necessary to re-cinch the saddle sometime later. Failure to re-cinch the saddle can produce a dangerous situation. The funniest incident concerning a loose cinch strap came in 1962 when two paleontologists came out from Washington, D.C., on a field excursion. D.J. and I decided to take them to a poorly exposed rock section in a remote area where fossil identification would help us determine the geologic structure.

We rented two horses for them to ride from an outfitter and within two hours of riding we reached the area. Then the fun began — the very tall paleontologist didn't know how to get off his horse! As he sat there, the saddle started to slip, very, very slowly at first. The paleontologist just sat there, his eyes getting bigger and bigger. He didn't say a word. Suddenly, the saddle slipped completely underneath the horse, with the paleontologist hanging on for dear life! It was all utterly priceless. We expressed our concern as we pulled the disheveled man off the saddle. Luckily, the horse was an old gelding and just stood there. We assured the fellow that the ride back would not be as eventful.

. . . .

The day started like it always did. In July, dawn breaks early. D.J. and I were heading up Palisade Canyon, a east to west breaking canyon in the Snake River Range. The lower reaches of the canyon are fairly wide open with very little pedestrian traffic. There is only a short distance that a truck pulling a horse trailer can travel. As we rode our horses up the canyon, we saw a rather strange sight. A small stream coming into the canyon had a group of boy scouts near it, grouped around one older boy. They paid no attention to us because the noise made by the rushing water in the main stream and partly because the trail was wide open and flat so we made no significant noise.

D.J. decided to stop and look at an outcrop while I continued to ride up the trail. Suddenly, my horse stopped. Slowly, he started moving over to the group of scouts. Well, I thought, let's see what the boys are up to. The horse slowly edged his way into the circle of boys, and then extended his head over the older boy's shoulder. The scout was holding a grouse, apparently one they had knocked out of a tree with a rock. The interest of the scouts seemed to be focused on how to clean and skin a grouse; with the older scout demonstrating the whole process. The horse must have been amazed because he leaned over the scout's right shoulder for several minutes. Still the scout rattled on. The horse then put his head within six inches of the grouse itself, taking a good look at the entire demonstration.

At this, the scout suddenly flinched. Whirling around

he looked up at me and gave a loud shriek, dropped the grouse and ran down the trail. The whole troop was right on his heels. As I sat there on my horse, it was obvious he thought I was the game warden. The fine for taking a game bird out of season could be quite high. It was also obvious that the boys wouldn't be coming back. As I sat there, I wondered if there was a penalty for picking up a dead bird — there shouldn't be any. I got off my horse and picked up the grouse and showed it to my horse. I said, "What do you think?" He seemed to nod his head in approval. I wrapped the grouse in a plastic bag and placed it in my saddle bag.

. . . .

The field season was rapidly coming to an end and I would be returning to Denver. Change was in the air and my personal life was suddenly going to change forever! However, I had a problem, I had a hard time sleeping at night because I couldn't forget the conversations I had overheard on the Nevada Test Site about the radioactive clouds that spread across the United States. The venting of nuclear explosions to the atmosphere was apparently a topic that was not supposed to be discussed publicly.

The dying of sheep in the Nevada desert and higher cancer rates in several western states was only part of a much bigger problem. No investigations have ever been made by the U.S. Government to follow the general track of radioactive clouds across the United States and correlate them with higher cancer rates. It is important to know if there is any correlation so the proper medical treatment can be determined. Cancer caused by toxic chemicals introduced into a water well, and cancer caused by radioactive fallout, both produce the same effect. Severe health problems that can go unrecognized for long periods of time, even in a small geographic area.

The fetish of trying to hide government bungling has been accentuated in 2012. The Justice Department had proposed a change to the Freedom of Information Act that would allow federal agencies to say that <u>certain records don't exist</u> when in actuality they <u>do exist</u>. Theoretically, this would take place only if national security was threatened. However, who actually makes this determination? The information about the serious spread of nuclear radiation over the central and northeastern United States is an example of the type of information that would be subject to be suppressed and even denied even though it may have caused serious health damage in the United States. The government <u>of the people, by the people, and for the people</u> apparently only exists for the politicians! This proposal for hiding government mistakes was eventually turned down, but it will ultimately surface again.

Chapter 6

THE ODYSSEY CONTINUES…

My fiancé and I had made dinner reservations at one of the more popular dining areas in Denver. It was the Christmas holiday season in 1962, and we needed to make plans for our spring wedding. We had met at one of the local single groups that dotted the Denver area in the early 1960s. The churches had established a number of gathering spots where singles could get acquainted. There was a single's bowling league, a single's square dance club, and single's supper clubs at the churches where dinners would be served on Sunday evenings. In 1962 there was no internet, no Facebook or Twitter, or e-Harmony or Christian Mingle sites. It was amazing, in order to connect with somebody you actually had to see and talk to them in person!

We hadn't known each other very long. We had said hello to each other once in the spring of 1962. Unfortunately, I had to immediately leave for my field area as my sojourn on the Nevada Test Site seriously delayed my return to my field area in southeast Idaho. On my return to

Denver in September we had met and she had invited me to be on her bowling team. Obviously, things had rapidly spiraled out of control from there and after three formal dates we became engaged. The field season would necessitate an early spring wedding. She had indicated that a lovely summer wedding would be more appropriate as we hadn't known each other very long. However, I had been hearing rumors that my field area was going to be changed. If it was going to change, I would need to be very involved in that planning and arrangements for a summer wedding would be very difficult. After very, very much discussion, a date was set; we would be married in a small brown-stoned Presbyterian church in early April in Lake Crystal, Minnesota, a small farming community.

My field area was changed that spring. I was now going to work in Teton Valley, Idaho, which was the Valley immediately west of the Grand Teton Range. I seemed to be getting closer to Jackson Hole, Wyoming, every year!

In 1963, the mapping program had started to slow down; attrition was starting to have an effect. Of the six geologists detailed to the Nevada Test Site, three had stayed there at the invitation of that group. Another geologist was killed in a tragic traffic accident in early January. He was a passenger in a small car that got hit on a mountain road west of Denver. Sitting in the passenger seat without seat belts, he was thrown through the front windshield of the car with the accident severing his head. Of the six senior geologists originally assigned to southeast Idaho and northwest Wyoming in 1960, four had either left the Conservation Division or were leaving by 1963.

I had expected to continue with the mapping program

in the Swan Valley area. However, the two geologists working on the Driggs 15-minute quadrangle which covered Teton Valley, had returned to the Geologic Division. Their work had never been completed. Consequently, I received the assignment to complete the geologic work and finalize the map for publication. When I received their initial maps showing their field work, I was very surprised. Obviously, I would have to spend a lot of effort in Teton Valley.

Teton Valley was formerly known in the fur trapping days of the early 1800s as Piere's Hole. Teton Pass was the main conduit for travel and trade for Indian tribes between Teton Valley and Jackson Hole, Wyoming, long before the fur trade started. John Colter was the first white man known to have traveled across Teton Pass into Teton Valley in 1808 on his way to discover Yellowstone. John Colter was born in 1775 in Virginia and ultimately became a typical woodsman like Davy Crockett. His grandfather had came from Ireland to America about the year 1700. In 1803 John Colter joined the Lewis and Clark expedition as a hunter and eventually left them in 1806. At that point he joined a brigade of 42 men headed by a fur trader called Manual Lisa who was interested in trading with the Indians. It was on that expedition that Colter discovered Jackson Hole, Teton Valley, and Yellowstone.

The Colter Stone, a small boulder of black volcanic rock, was found in the eastern part of Teton Valley in 1931. It was about the size of a human head with a flatten side that had the inscription J COLTER 1808 carved on it. The discovery of this stone seems to have cemented Colter's place in history. From that time forward, Teton Valley became a gathering point for a rendezvous between fur traders and various Indian tribes. In 1832 a rendezvous in Teton Valley led to a fight between the mountain men and Indians of the Blackfeet tribe, with a number of Indians killed. After 1840 no more rendezvous meetings were held as beaver hats went out of fashion.

Teton Valley, as well as Jackson Hole, was never part of the Louisiana Purchase because it lies west of the continental divide and was a part of the Columbia River system. This entire area was known as the Oregon Country. John Colter joins a long list of mountain men who traveled through this general area, men such as James Bridger, David E. Jackson, Jededial Smith, Seth Grant, William Sublette and others.

Newly married, it was imperative that my wife and I obtain adequate housing for the summer months in Teton Valley. The main reason one of the previous geologists had left was because he could find no housing for his family. Consequently, he had put up a tent for his wife and small twin boys near Pine Creek Pass west of Victor, which resulted in his wife having a nervous breakdown. Fortunately, my wife and I found an apartment for rent over the Post Office in the town of Driggs, Idaho.

The apartment over the Post Office proved to be adequate for our needs. There were three apartments on the second floor, with all three reached by a single outside stairway. The entire building was heated by steam heat. The laundry on the second floor consisted of a wringer type washing machine. After washing clothes, the clothes had to be run through rollers to remove the soapy water, then rinsed and run through the rollers again. The clothes dryer consisted of clothes lines in back of the Post Office, which necessitated hauling a heavy clothes basket up and down the outside steps.

In 1963, the town of Victor had a population of only 300 residents, and Driggs had a population of around 800 residents. The entire valley was predominantly of the Mormon faith and the majority of the people had been no further than Idaho Falls, Idaho, about 50 miles away. My wife and I had only three of the local residents who would talk to us as we were "outsiders." One of the residents, of course, was our landlord. The second was one of the local grocers who wanted our business. The third was a local rancher who I rented horses from and who also did some packing for me into the back country. My wife felt very isolated in this type of community. Prior to our marriage she had been very active as a Registered Nurse working in the Denver School System. Now, she was occupying her time by watching "soaps" on the TV set. I am sure this living arrangement was not what she had envisioned when we got married!

There was some entertainment available as the town had a Drive-in or Outdoor movie theatre. Once a month we escaped the valley by driving to Idaho Falls for grocery shopping. On one of our first weekends, we drove past Victor and stopped at the local restaurant for breakfast. In looking at the menu, I noticed that liver and onions were only 65 cents. "Boy," I told my wife, "that's a good deal." My wife agreed that it sounded very good. After I had ordered, the waitress suddenly turned up again very red-faced and grabbed the menu and gave me a different one — but now the liver and onions were 85 cents. Apparently, I had been given the "local" menu by mistake.

The town of Victor did have one thing going for it and

that was a melodrama given on weekends. One could boo and hiss the villain all during the show — and in the meantime throw peanut shells on the floor. My wife and I tried to attend the show several times each summer. Relatives from Minnesota came to Driggs the first summer and we took them to the melodrama. They enjoyed the show so much that they talked about it for years later.

Midway through our first summer, we heard about an auto accident near Petterson's store in Irwin, near where I had stayed the previous two summers. We couldn't believe what we had heard and had to go to Swan Valley to see if it was true. A Dodge convertible, traveling at a very high rate of speed, had slid off the road near Irwin at 2:00 a.m. It had traveled over 300 feet sideways and had just nicked the gas pumps in front of Petterson's store. The car continued on across a gravel road next to the store and hit a log cabin, knocking the side of the cabin inward.

The aftermath of a high-speed automobile accident that would have turned tragic if the car had not narrowly missed the gas pumps of Petterson's store.

The car ended up in the middle of the cabin with the roof falling down over the car and hiding it from view. Ernie Petterson had crawled under the roof on his hands and knees with his flashlight to find the driver alive and without a scratch — and slightly drunk! Fortunately, the occupant of the log cabin had spent the night in Idaho Falls. The accident could have been more tragic if the gas pumps had been hit. These pumps were connected to a free standing 2,000 gallon gasoline tank on the south side of the store. Rumors later surfaced that the driver had been bragging that he could hit 140 mph with his convertible!

The field mapping project covered both sides of Teton Valley and was not as rugged as the areas I had been working. Most of the areas I could reach with a jeep. Horses, however, would still be very useful in certain areas. I needed a horse trailer and went to Jackson where I picked up an old, open, wood horse trailer from another geologist. The trailer had seen much better days but I calculated that my usage would not be that great. In the end, I did use that horse trailer quite frequently. Unfortunately, the end came much sooner than I expected.

One Monday morning I picked up two horses from the ranch and started traveling on a gravel road I had used before. Over the weekend, the county had graded the gravel road and had left a gravel ridge in the middle. With no warning signs posted, I hit the gravel ridge. The horse trailer jackknifed, forcing the Jeep and the trailer off the road. The trailer tipped over completely, throwing both horses out of the trailer, and then righted itself. One horse was still tied by his halter rope to the inside of the trailer and had his neck slightly stretched, but did not seem to be hurt.

The other horse had slid across the gravel road on his back. Unfortunately, he did it on my brand new Heiser saddle. Both horses survived, apparently in good shape with only small patches of skin missing here and there.

The horse trailer was made completely of wood and had lost a number of boards. I gathered what we could find and pulled the trailer back to Driggs. I nailed most of the boards back on the trailer, and it really didn't look too bad. The axle was a different story. It was bent so badly downward that the entire trailer resembled a pregnant goose trying to lay an egg. I hauled the trailer over to the local blacksmith. Unfortunately, even he could not get the bend completely out of the axle. I took the trailer back to Jackson and explained to the geologist's wife that I didn't need it anymore and left as quickly as I could.

In 1964, I returned to Driggs to continue working on the geology in the area. This time it was a little different than the previous year. Now there was three in the family, my daughter Julie was born in May. We had also purchased our first house, which as time went on would be our major home for the next 50 years. My project area had now been moved to the Rendezvous Peak quadrangle, an area immediately east of Victor, which was the very southern part of the Grand Teton Range. The main resource of interest was phosphate, which occurred throughout that quadrangle. The ski tram build at Teton Village on the Jackson Hole side rises to the top at the very northeast corner of that quadrangle. I was working in the area of Teton Village when three tram workers died when a wire broke as the men were trying to ride a cement bucket down to the tram station.

At the time I worked in Teton Valley, that valley seemed

to be a peaceful interlude between two entirely different geologic environments. The vertical forces that formed the Teton Range are characteristic of the continental shield, i.e., the old structural platform that forms most of the continent between the Rocky Mountains and the Appalachians in the east. The average tourist in Jackson Hole stands in awe at the magnificent beauty of the Teton Range. The Teton Range is an upthrown block along a geologic fault, a fracture in the earth's crust along which movement has occurred. In the case of the Teton Range, estimates of movement along the Teton Fault range from 18,000 to 35,000 feet. As you can guess, geologists sometimes have a lot of indecision about their scientific analyses! The western side of Teton Valley is part of the Idaho-Wyoming Overthrust Belt, in which the dominate force is horizontal. It is the collision between these two types of structural plate boundaries that form a great potential danger to all those who live in Teton Valley and the Jackson Hole area of Wyoming. As always, the problem is Yellowstone.

Chapter 7

THE LOPSIDED DEER OF
TETON COUNTY

My uncle has seen them many times

The Teton Range is the most photographed mountain range in North America. In the years that I worked in the Jackson Hole area, I constantly ran into tourists coming from Yellowstone Park who commented that the Tetons were much more beautiful than anything they had seen in Yellowstone. Yellowstone does have an abundance of wildlife however, but frequently a tourist has to travel or hike into remote areas to observe them. In 1962, a young male tourist with my name was killed in Yellowstone by a rogue buffalo when the tourist got too close to the buffalo when taking a photograph. I received several calls from other geologists wanting to see if I was still alive!

The Rendezvous Peak quadrangle I was mapping in 1964–65 was a continuation of the geology mapped in Teton Valley. The topography was one of sharp valleys and high ridges and the geology could only be worked on horseback along with a lot of camping. The quadrangle covered only the very southern end of the Teton Range, extending only seven miles north of Teton Pass. The ski lift at Teton Village is near the northeast boundary. The geology of the area was relatively simple, mainly westward dipping rock formations cut by numerous high-angle faults.

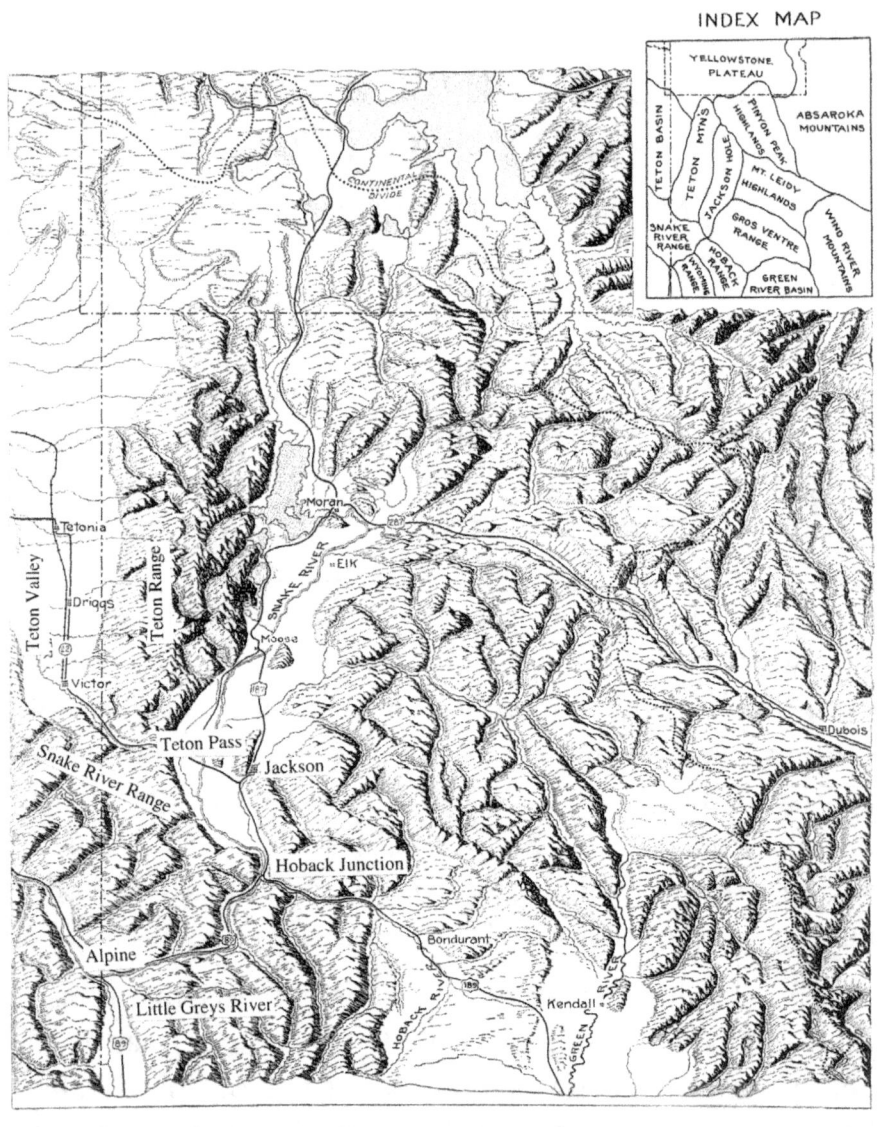

Relief map of Jackson Hole, Wyoming and adjacent areas
by S.H. Knight — 1956, Wyoming
Geological Association Guidebook, 15th Annual Field Conference, 1960

These high mountain peaks are held up mainly by the Madison Limestone

Little did I know at the time that the seemingly geologic simplicity was eventually going to be much more complicated than it seemed. Plate 1 shows the published geologic map of the area.

I could tell many stories about my first summer in the Tetons, but there is one I remember more than the others.

. . . .

The campfire started throwing off more sparks and flamed much higher as the kid threw more wood on the fire. He was standing and waving his arms in the air and was literally yelling, "I tell you, I know it for a true fact, there are lopsided deer out there. My uncle has seen them many times!"

Typical horse camp in one of the higher valleys of the Teton Range.

Sitting across the campfire from him that last night as the week drew to a close, I could only sigh. There was one thing that old miners and old ranchers had in common; they would spin tall tales to anybody that might believe them. I had tried arguing with the kid, but it was useless. His uncle and father had both tried to convince him that each one was right and that the other one was wrong. His uncle had promised him $100 if he could take a picture of such a deer and prove that his father was wrong.

The whole situation was starting to get out of hand and serious. Today the kid had literally sailed 20 feet off his horse as he yelled, "I think I see one." He threw himself on the ground trying to rapidly adjust his 20-power telescope as he tried to focus on a rapidly disappearing deer across the canyon. His horse spun around and started running back down the trail. It took about 15 to 20 minutes for me to finally corral him as the horse finally had slowed down to a fast walk. A fork in the trail had enabled me to finally get in front of him, but then the horse was still a little spooked.

I had my doubts about the 18-year-old kid from the very beginning. My wife had taken my young daughter back to Minnesota as I was going to be camping for several weeks. My field assistant had become ill and a local rancher had persuaded me that his nephew could then help me. The kid had just arrived from New York and wanted to experience the Wild West. I said, "Can he ride a horse?" The old rancher wiped his brow as he stood holding his hat. Finally, he said "Well, that probably is one thing that he is actually capable of doing!"

The American West abounds with tall tales. Mark Twain started his story telling by quoting stories that the old miners told him while he was in the gold fields of California in the 1860s. They would sit around the campfires every night and each miner would try to top the last story that was told. Mark Twain achieved some fame when he told the story of, "The Celebrated Jumping Frog of Calveras County," in one of his newspapers columns. The story concerned two miners who bet their Bull Frog could out jump the other frog. The story went on to show how the competitor won the match by catching the Champion Bull Frog one night and pouring lead shot down its gullet so he couldn't jump during the competition. While that made a good story, the story the old rancher was telling his nephew was almost as good. "It seems," the old rancher said, "that some deer become lopsided because they grow legs a couple of feet longer on one side so they can run faster around the steep mountainsides and escape the hunters!"

The letter from New York had arrived late that October morning and had been sitting on my desk most of the day. The handwriting on the envelope had been short and brief, "Mr. Schroeder," it said, "Denver Federal Center, USA." The writing was unruly with big letters, obviously written in haste. I was surprised it had gotten to me with that address. I already had a certain premonition what it was going to say in the letter. Meetings had been taking up most of my day but the time had come when I needed to open it.

Inside the envelope was a short letter, a note really. It said, "I told you it was true." There was a photograph of what looked like a lopsided deer, not the best picture, but one that certainly looked very authentic. I had to smile as I thought, "Who would have sent this picture to that 18-year-old kid." As I filed the picture in my desk drawer, I looked out the office window as the snowflakes started to increase in intensity. Sometimes imagination had to give way to reality; I was sure the kid would love getting the $100 from his uncle!

Chapter 8

JACKSON HOLE DAYS

The sword of Damocles hangs by a thread over Jackson

In the early summer of 1965, I stood on a high mountain peak in the southern part of the Grand Tetons that overlooked Jackson Hole. It was hard to realize in 1965 that only 157 years had passed since John Colter became the first white man to enter the valley. Archeological digs had shown that the area had been popular with many Indian tribes long before the fur trade developed. Among the Indian tribes coming to the valley were the Crows, Banocks, Shoshones and the Blackfeet among others, all wanting to kill and dry wild game for the long winter months.

There were many fur companies that conducted business in the valley. The American, Rocky Mountain, Pacific, Hudson's Bay and Northwest companies were all familiar names in the history of the region. Along with these companies were individual trappers willing to sell to the highest bidder, as well as trappers who would trap on shares with one specific company. Companies would form, split up, and then reform with different individuals involved in the business. William Sublette named the valley "Jackson Hole" in 1829 after David E. Jackson, another hunter and fur trapper who loved the valley.

When I started working in Jackson Hole, the first question I had to answer was simply, "Where do I stay with my family?" The town of Jackson was expensive when it came to finding housing for the summer. It is even more of a problem today because the town has become a tourist mecca. My answer in 1965 was to buy an old 1950 Anderson trailer and haul it to Jackson with a government truck. It has been owned by an elderly woman who lived in northeast Denver. At first, we stayed in a trailer court where the Virginian Motel now stands. Later we moved into the B&B trailer court on the south edge of Jackson. With just one child, the old trailer was adequate for our summers. We had placed a play pen in the front of the trailer for our

My 1950 30-foot Anderson trailer in a trailer park south of Jackson, Wyoming

baby daughter to sleep in. When she started to wake up and cry in the early morning hours, we finally hit on the trick of placing crackers on the counter next to her bed. Occasionally, when we heard her wake up early, we would see one little hand reach over the counter to grab a cracker and then complete silence.

In 1965, Jackson was a sleepy western town that was just starting to wake up. The Pink Garter Theatre was a very small theatre just down the street from the corner drug store. We bought several of Ken Faye's oil paintings of the Tetons from his Happy Peasant art gallery, which we still have hanging in our Denver home. The western cowboy shootouts on the town square always produced a big crowd. The Wort Hotel had wonderful evening shows. The restaurants had excellent food, heavy on prime rib as befits a western town. It was hard to believe, but we could actually go in the bank and cash a personal check. The tourists seemed to be taking countless pictures of the elk horn displays that framed the town square. In those years of the 1960s time seemed to stand still.

Unfortunately, the boom in tourist traffic in recent years has produced major traffic jams. Driving my motor home on my return from Alaska in 2008, I stopped in Jackson to see the 4th of July parade. The contrast in the area from the 1960s to 2008 was very dramatic. Homes costing over a million dollars now dotted the entire region. Teton Valley, with the towns of Driggs and Victor, had now become a bedroom community to Jackson. The general atmosphere in Jackson was one of complete satisfaction for all those who could afford to live there. It was several years later when I was reviewing the tectonic structure of the area that

I realized Jackson and the other towns sat over a major fault which extends all the way to Yellowstone. When Yellowstone becomes active with earthquake activity, these towns will start to exist like the fabled story of the Sword of Damocles. In that old story from mythology, the courtier had a single human hair holding the sword over his head — I'm afraid Jackson won't be so lucky! The potential for destructive earthquake damage could be overwhelming in its intensity!

When I started working in Jackson Hole, I knew that I was going to encounter many problems. Misconceptions had gotten into the literature because no in-depth studies had been made. Various geologists would wander into the valley, develop their theories, publish an article or two that was suppose to solve all the geologic problems. The geology of Jackson Hole is like the story of the five blind men who all touched a different part of the elephant. Unless you experience the entire beast, your understanding will suffer greatly.

When an individual looks at the Grand Teton Range, he doesn't have to be a rocket scientist to realize that there are tremendous vertical forces at work to create this mountain range. Surprisingly the Teton fault, which created this mountain range, doesn't present danger for earthquake damage to Jackson. The real danger is a geologic fault that has been misinterpreted in the past as a thrust fault. The fault known as the Cache Creek thrust fault is actually a high-angle to vertical fault which forms the terminus for the Teton fault. The amount of vertical movement along the Cache Creek fault is at least as great as the movement along the Teton fault.

The Cache Creek fault extends northward from Teton Pass through Teton Valley and is last seen disappearing under elements of the Snake River Plain. To the south, the fault extends eastward underneath the town of Jackson and continues southeastward along the north side of Cache Creek. It projects through the northeast corner of the Bull Creek quadrangle and continues southward along the Gros Ventre range. The horizontal length of this major fault can be shown to be a minimum distance of 50 miles, with a projected overall length of at least 75 to 80 miles.

The danger that this fault presents cannot be overstated. When the magma chamber beneath Yellowstone begins to fill, it can cause shock waves that will travel along the Cache Creek fault shear zone. The shock wave could be extremely severe and cause massive earthquake damage, especially to a town like Jackson which is located on top of the fault. Seismic stations need to be placed on the northern end of Teton Valley near Driggs, and possibly along Cache Creek, to monitor earthquake potential. Earthquake insurance of a substantial amount should be considered if there is any indication of danger. Geologists believe that the Yellowstone area has suffered three eruption cycles, with the last one about 640,000 years ago. The Yellowstone area today is rising about 0.6-inch per year, indicating that the next eruption of Yellowstone may not be too far off in the future.

The Cache Creek fault is unique in that it hides in almost plain view, with only occasional small exposures of its existence. In retrospect, I could have determined the true nature of this fault over 40 years ago. I had started mapping the Cache Creek quadrangle east of Jackson when I was told by Survey personnel that I had to move to another area. Apparently, I had started mapping over some student from the University of Wyoming who was engaged in mapping part of the area for his Master's Degree in geology.

The fault pattern at Teton Pass involving Paleozoic rocks indicates lateral movement, i.e., one structural element moving in a horizontal direction past another. The very poor exposures of the Cache Creek fault in all areas make this determination almost impossible to detect outside of the Teton Pass area. Even here, the horizontal movement has been basically hidden by the eastward moving Jackson thrust fault which seems to partly override the trace of the Cache Creek vertical fault.

The town of Jackson faces earthquake hazards in several ways. First is a shock wave coming from Yellowstone that could cause extensive damage. The second is recurrent movement along the Cache Creek fault as it slides past the Teton Range. In either scenario, the potential for damage could be quite extensive. When I originally mapped the geology of Teton Pass in the late 1960s, I felt I was dealing with a vertical force field of some type. When I drew cross-sections through Teton Pass prior to publication, I was quickly descended upon by several "senior" geologists from the Geologic Division. They insisted that I show the geology as a "thrust type" fault which would indicate major horizontal movement as that was the previous interpretation.

. . . .

"The doctor didn't come in today," my wife proclaimed as she came into our old Anderson trailer. I remember looking at her that day and saying, "What doctor?" "Don't

80

you remember," she said, "I told you about this famous surgeon from the East Coast who gave up his lucrative private practice to come to Jackson?" I had to acknowledge that I remembered something about it; something about her meeting the doctor. I had been compiling my geologic maps and whether some doctor managed to go to the hospital on time was the least of my current concerns. The doctor not coming to work one day suddenly became the topic of conversation throughout the valley when nobody could find the doctor — or his wife. The front page of the Jackson Hole Guide newspaper gave a weekly summary of the continuing search.

Initially, nothing was released by the authorities about the disappearance. Finally, the news broke about the doctor and his wife being murdered in their home. As the news about the murder continued to spread, new details kept emerging. The log house owned by the doctor and his wife lay at the foot of Teton Pass, and I drove past it every day on my way to Mosquito Creek. The road up this valley provided the main access into the Teton Pass quadrangle. The log house was located only a small distance off the gravel road I was using and it was easily visible in a small grove of trees.

Authorities had finally broken into the home and found a very disturbing scene in one of the bedrooms. Bullet holes and blood stained walls indicated there was no doubt a homicide had occurred. The immediate suspect was the couple's son. Deputies followed him wherever he went, but he was not arrested because no bodies were found. It was rumored he had some high gambling debts in Las Vegas. That fall, an undertaker from California was hunting up

Mosquito Creek; he stopped suddenly when the wind gave him indications about a nearby trash dump. The doctor and his wife were both found in a shallow grave. The son was subsequently arrested and convicted of the murder of his parents. The trash dump was only about 30 feet off the gravel road I was using every day along Mosquito Creek.

. . . .

I moved my old 30-foot trailer to Bondurant when my field area changed from near Jackson to the eastern end of Hoback Canyon. This spread-out town lies in a very open area with breathtaking views of the surrounding mountainous areas. I had located a very small trailer court that looked to be ideal for the summer. We had been there for several weeks when a problem developed. The trailer court was in a low area and the sewer started to back up, placing water drainage over a large portion of the park. When my wife complained, the woman running the court said she couldn't do anything about it — and if we didn't like it, we could move. We had no choice but to move, back down the canyon to Hoback Junction which is located 14 miles south of Jackson.

Hoback Junction was famous during the era of the mountain men. The Hoback River runs out of the Hoback Canyon to join the Snake River. A fur trader by the name of Wilson Hunt named the Hoback Canyon in 1811 after one of his guides who loved trapping in the canyon. Many mountain men traveled through Hoback Canyon on their way to a rendezvous with Indian tribes near the headwaters of the Green River. A local rancher around the year 1900 claimed to have found the barrel of a flintlock rifle made in London in 1776 in a small canyon off the Hoback River. Historically, we can say that the Hoback Canyon was in continuous use by the mountain men ever since fur trapping started in the early 1800s.

The move from Jackson to Bondurant, then to Hoback Junction, was only a minor problem. I had much bigger ones. Before I left Denver that spring, I had a meeting with one of the senior geologists in the Geologic Division about the upcoming field season. His main comment was, "Marvin, the age of the Paleocene Hoback formation is based on vertebrate fossils, yet the oil and gas well drilled in the Hoback on Sandy Marshall Creek has 8,000 feet of Cretaceous pollen. How do you do that?" It was a simple question that ultimately had a simple answer — somebody had misinterpreted the geologic data.

Even before I started working in the eastern part of the Hoback Canyon area, I had the feeling that the mapping effort was going to be difficult. I had attended a small conference convened by the oil and gas industry in the Virginian Motel in Jackson to discuss various interpretations of the Hoback Basin. As I looked at five different geologic maps laid out on a conference table, I just stood there in complete amazement. All five maps were different, no two maps were the same. These different maps helped convince me that these various interpretations needed to be solved once and for all time. The two quadrangles I mapped in that area were the most difficult projects I had in my geologic career. I rode over 1,000 miles on horseback and also wore out a pair of vibram-soled field boots every year while I tried to determine the complexity of faulting involved in the frontal zone of the Idaho-Wyoming Overthrust Belt.

The answer to the question about rock identification was quite simple after I had mapped the area. A geology student from Camp Davis, the University of Michigan field camp, had collected vertebrate fossils in the 1950s from a landslide of Paleocene age rocks that had slid over Cretaceous rocks. In addition, he didn't recognize that a major fault also was present. Geology has never been known to be an exact science, it is only our interpretation that will change with time. As for the geology student, he went on to become Chairman of the Geology Department at the University of Michigan — and I must add, an outstanding geologist!

Hoback Junction was not the place for a young family. My trailer backed up to a steep drop-off of over 40 feet to the Snake River. My wife never let our young daughter leave the trailer by herself during the time we were in the park. Fortunately, the front office had bought me a new "state of the art" steel horse trailer after the debacle with the wooden horse trailer. The logistics of hauling horses up and down Hoback Canyon was much easier. In the end, I hobbled horses in my field areas whenever it was possible. Near the end of the field season, we moved our trailer to the B&B trailer court in Jackson.

In late summer of 1968, I came close to suffering a serious injury. I decided to take two horses back to Swan Valley from Jackson, before trying to climb the Grand Teton with another geologist. He came to Jackson in September and had persuaded me we could climb to the saddle between the Middle and Grand Teton in the afternoon and camp next to one of the guide cabins. When the guide and his client appeared at 5:00 a.m. in the morning, we would simply follow them and use them as our guide. Unfortunately, when I took the horses to Swan Valley one of them kicked me, just grazing my left knee. If the kick had handed squarely, it would have broken the entire knee joint. The climb to the top of the Grand Teton never happened as I limped around for a week.

Much later, when I told this story to an instructor at Camp Davis, he laughed and proceeded to tell me his story about the Grand Teton. He had decided to try and climb it when the field camp was not in session. As he was backpacking up the trail in the early morning hours, he heard a lot of noise behind him. Jumping off the trail he saw a small group of men, short and as wide as a barn door, running up the trail. Late that afternoon while he was still climbing, he heard the same noise again and once more jumped off the trail. The same group of men came running back down the trail after climbing the Grand Teton; later it was learned they were a group of Sherpas in this country from Nepal and were used by climbers on Mount Everest.

I had major physical problems in 1968 with two horses I used while mapping the geology around Camp Davis. South of that camp a dirt road extends west off the main highway over a ridge to a dude ranch. I came to the conclusion that the easiest way to map the surrounding area was to simply rent horses from the dude ranch. The dude ranch rented me two horses — one a fairly large palomino, and the other a beautiful spirited black horse. Field work soon turned out to be impossible. The palomino lost significant weight and became very unstable and tended to trip over anything that was rough ground. The black horse, which I rode, had soft hooves and constantly threw horse shoes with the result that he developed tender feet and started to limp badly. I took the two horses back to the dude ranch and said, "Sorry, I can't use these horses." The dude ranch now filed suit against the U.S. Government, claiming that I had abused and ruined their horses. The solicitor settled the claim for $300 much to my dismay. Later, I learned that the dude ranch had obtained those two horses from a horse trader in southeast Idaho.

The trouble with the dude ranch continued. I kept on using the dirt road as it was on government land and gave me access to my field area. I usually pulled the trailer and horses to the top of the ridge and parked there. One day in late August I returned to the rig and found a notice of trespass on the windshield. I ignored it! When I returned to Denver in the fall, I received a letter from the local magistrate in Jackson, Wyoming, ordering me to turn up in court or he would issue a summons requiring my appearance.

I went to the solicitor's office at the Denver Federal Center and met with his assistant. He pulled out one of his law books that lined his entire wall and looked up the U.S. Code. He said, "I would like you to write a letter to this magistrate and request a transfer of this trespass to the U.S. Magistrate in Cheyenne, Wyoming, as provided for in the U.S. Code. When you are notified it has reached there, I will try to see what I can do on your behalf." In the end, I signed a letter stating that I would not trespass in the future. Theoretically, the road was a private road because the dude ranch maintained it even though it was on government land.

. . . .

As he stood in the door of the cabin, he was one of the most deplorable-looking field assistants I ever had work for me. He was unshaven, hunched over in his old worn bathrobe, with matted hair, puffy cheeks and bleary red eyes squinting against the morning sun. "Mr. Schroeder," he lisped, "my contacts scratched my eyeballs and I need to spend the next 24 hours flat on my back with my eyes closed."

Quite obviously, he wasn't going to do any work today. I had driven my government jeep from Jackson, Wyoming and usually picked him up early in the morning on my way to my field area. Today was going to be different. The U.S. Geological Survey had always insisted that when you work in remote areas you absolutely had to have an assistant with you for safety reasons. Fortunately, I had additional air photos with me for a different field area where I could probably work by myself. I tried, but that area was still too rugged for working alone.

So after careful consideration, I decided to return to Jackson to finish compiling some of my maps. Driving by Hoback Junction, I decided I needed to fill my gas tank. Pulling into the station I stopped and stared in utter disbelief — there was my field assistant pumping gas. He looked absolutely fine, hair neatly combed, freshly washed and ironed blue jeans, and no red bleary eyes, just the picture of good health. However, it was now too late to go into the area in which I need to work. So I smiled. As I was leaving, I told him I was glad to see him feeling better.

By early afternoon I finished compiling my maps and decided it was time to have a positive discussion with my field assistant. When I arrived at the cabin, he was nowhere to be found. The door was hanging open and everything was gone — all the clothes, books, pictures — the cabin had been stripped clean. Fortunately, he did not have any of my work materials, such as field notes, maps, etc. Since it was early September, the office would not be hiring another assistant at this late date. I would have to return to Denver.

I had problems with this assistant in the past. We had been out in horse camp several weeks ago and had spent part of a day making a long traverse up a steep mountainside. It had rained the night before and the slope was muddy and slippery. I had lagged behind him while making some field notes. Suddenly, I heard some yelling and screaming up ahead. I grabbed the reins of my horse and literally pulled him up the hill.

The scene was a rather pathetic site. The kid had tried to lead his horse across a small downed tree at an angle on this slippery slope instead of taking the extra time and leading the horse around it. The horse has slipped and fallen with two legs beneath the log, with his belly resting against the log. All attempts by the horse to get loose failed with the slippery slope. "What are we going to do" wailed my assistant. "Yes," I thought, "what are we going to do?"

There were axes back in camp, but it would be difficult to find this exact location again since it was in a remote area. In addition, the longer the horse was in that position he could be badly hurt trying to get free. It could take a week to find another horse, and we would be stuck paying for this one. If we could not get the horse loose, we would have to haul the saddle and field gear to camp. In a wild moment, I considered the possibility of putting the saddle on my field assistant.

I removed the saddle from the horse, which seemed to calm him. Fortunately, the tree had just come down, so my rock hammer easily splintered a little of the wood. "Get to work" I told my assistant. "We are going to hack off a little wood at a time" — it worked! About an hour later we had removed enough wood so we could move the top of the tree; with considerable effort by the horse, he was also free. We returned to camp disappointed, it had been a wasted day.

Arriving back in camp, I found the two Mormon boys I used for packing in a very gleeful mood. "We are going to have fresh meat for dinner tonight. We shot two grouse." Of course, I knew that they didn't have any guns. Their fun was with the field assistant as they described how good the stew was going to be — much better than canned meat. Finally, they got the assistant to eat most of the stew. At that point, they rolled on the ground with much laughter. Apparently, they had killed an old rabbit with big rocks and passed it off as fresh grouse.

I had about enough for the week. I knew that one of the Mormon boys had his one year wedding anniversary today. I looked at the sky — more rain coming. The horse had a pronounced limp. I walked around for a few minutes and finally told the boys, "Let's pull the camp, we are going in." In 30 minutes the horses were packed and we were heading out. As we left that late Friday afternoon, I thought, "Rocks are rocks, but wedding anniversaries are more important."

Now later, my assistant was gone. That night I put myself to sleep by thinking about other field assistants I had in the past. There was one assistant who came back to work on Monday morning quite giddy. He had been working in the kitchen of a local dude ranch over the weekend. Saturday night a group of college kids from Idaho Falls came into the bar area. Quite soon, after much drinking, the boys persuaded one of the co-eds to dance naked on the bar top by passing the hat. The assistant said she did quite well. Then there was a foreign geologist I had for six weeks, who would get off his horse and roll in the flowers. He said they didn't have fields of flowers in his country. For some unaccountable reason, he would occasionally sleep in the park in Jackson, next to the elk horn displays. One night he got propositioned by one of the older local ladies, who said she wanted to experience foreign culture… then the assistant who complained incessantly that his girlfriend would always forget to bring c..don… and then the one who had his dog shot by the Sheriff of Teton County for chasing cattle… and then…

. . . .

The car screeched to a stop by the horse trailer. My wife had caught me at Hoback Junction to give me the bad news. She had just put out a fire in our old Anderson trailer. The hot water heater cord had shorted out, producing flames and noise underneath the kitchen sink. She had ran out of the trailer and pulled the trailer's electrical cord from the park's electrical box. Back in the trailer, she found one end of the hot water heater cord still in flames, so she had stuck that end in a pail of water. Everything was now safe, but it was obvious that we needed a different trailer as this could have been a very serious incident.

The old Anderson trailer had served us very well. My son was born in 1966, so I had built bunks in the back end of

My family practices being "a geologist" one weekend

the trailer for our children. I put in an old hide-a-bed in the front that we slept in. I bought a used 21-inch black-and-white TV in Jackson to help keep my children occupied. I had turned the TV on one early morning to hear the news and was shocked to hear that Bobby Kennedy had been assassinated in California. However, the old trailer now had to be sold, and I received the same amount that I had paid for it five years before. I replaced it with a brand new 1970 Kit 22 foot trailer that had an actual bathtub in it. I would not worry about the new trailer freezing up every winter like the Anderson as I hauled the Kit trailer back to Denver every fall.

. . . .

Teton Range in the far background

Field work in and around Jackson Hole was starting to come to an end and I was going to move to a new field area. The new area wasn't that far from where I had been working, only on the other side of the Hoback Range. To get to my new field area, I had to travel 23 miles southwest from Hoback Junction to Alpine, Wyoming. From there it was about 18 miles back east on the Greys River road on the south side of the Snake River. This new area was entirely different from the Jackson environment, with no tourists, road congestion, or traffic jams. It was very remote with no development of any kind.

Our camp site was at the very end of the Little Greys road which had now junctioned with the main Greys River road. We had an idyllic camp site next to a small lake. The main lake, called Water Dog, lay several hundred yards further up a very narrow, twisting, and badly rutted dirt road on which only a Jeep could travel. That lake had a small hunting camp on its northern end, and also very nice looking rainbow trout in it. The little lake where we camped was more or less seasonal, but it did have water in it the entire time we camped near it. One year we shot off fireworks over the lake on the 4th of July.

The trip to our camp along the Greys River road was always an adventure due to steep drop-offs along a very dusty, narrow, and winding gravel road. There were many places where it was impossible to back up or get off the road when meeting another vehicle. The Conservation Division had acquired two Airstream trailers and I used one of them. I would tow the Airstream with a government truck, and my wife would tow our smaller trailer with our station wagon. The Airstream was a very nice trailer, but it had one major fault, it would soak up the dust along this gravel road.

Our third child was born in the fall of 1970, but with two trailers we had no problem with sleeping accommodations in the Greys River area. We had sold our Kit trailer and bought an 18-foot Great Divide trailer which was more amendable for back road travel. A Forest Service cabin was located less than a mile from our camp and had a fresh water spring. Every three days we filled 10 gallon milk cans at the spring site and used them to fill our trailer tanks with fresh water. We found it was necessary to haul our trailers back to Alpine every 10 days to dump our holding tanks, and get more groceries, propane, and other supplies.

The area I had picked out for our camp was ideal in that it was almost in the middle of the area I was going to be mapping. There was a nearby corral which I could use for my horses. One night they managed to break out of the corral during a thunder and lightning storm. After spending most of the morning trying to round them up, I finally gave up. Luckily, a sheep herder from Spain was nearby, and after I waved some dollar bills, he managed to use his horse to round up ours.

The area immediately coming into the camp site was wide open range, with plenty of meadow grass. A building several miles from our camp was apparently used by the Cattle Association as a "watch" facility to detect any potential cattle rustlers. The only visitors we had in the years we camped in that area were several carloads of college students who came one weekend to play music and drink beer by the upper lake. Unfortunately, their car radios ran down their car batteries and they were stranded the next morning. My survey of the lakeshore indicated they had been mainly involved in other, more interesting activities!

. . . .

I heard the dog growl before I heard the scream. My 6-year-old son had put his arms around my field assistant's malamute dog, and the dog had turned around and bit him in the face. The bite produced lacerations around his left eye and temple area. My wife worried that he would need stitches, and maybe even lose his eye sight. After much discussion, she decided to take him to the Jackson Hospital, a drive of at least 50 miles. So at 8:00 p.m., I stayed with our other two children while she drove to Jackson with my son, along with my field assistant. She still remembers, over 40 years later, returning to camp along that very narrow, winding Greys River road in utter darkness at 2:00 a.m. in the morning. Fortunately, the bite produced no lasting damage to my son's eyesight.

After the dog incident, I realized that I had been lucky not to have any serious injuries due to my fieldwork. However, I did have two serious non-field incidents. The first occurred after I returned to Swan Valley in 1962 from the Test Site detail. Several weeks after returning, I came down with extreme fatigue, nausea, blinding headaches, chills, and cold sweats; the bed was almost soaking wet in the mornings from my night time sweating. I could only get out of bed several times each day for a few minutes. After 10 days, I managed to drive to Idaho Falls to seek medical attention. I thought I had developed tick fever, but blood tests were negative. The medical feeling now is that I may have had a low grade reaction to radioactive poisoning obtained at the Project Gnome site in southern New Mexico, but that is entirely speculation.

The second incident occurred in 1968 when I developed symptoms of an ulcer one weekend while we were in Yellowstone. My wife drove me back to the Jackson Hospital, but a doctor was not immediately available for several days. We then decided to drive back to Denver the next day. Unfortunately, by that time I had taken too many pain pills and I could not drive. My wife drove me and our two children the entire 550 miles back to Denver in ten hours. During the drive to Denver, I kept trying to remember the words to that old western song, "Thank God she's a country girl!"

When I sat around the campfire on some of those long summer nights, I knew that this way of life was not going to last forever. I remembered the old story about a King who summoned his Wise Man one night and told him to tell him a story the next morning that was always true. "In fact," the King said, "if you do not tell me something that is always true, I will cut your head off!" The next morning the King summoned the Wise Man again and demanded the eternal truth. The Wise Man looked at him and said, "And this too shall pass away." This ancient saying applied to me too, and my next field area was going to be in southwest Wyoming where strange things were going to start happening.

Chapter 9

HORSES — HORSES — HORSES

My sojourn among the horse population in the Jackson Hole area was starting to come to an end by the middle 1970s as my field project area had been shifted to southwest Wyoming. My new horse was a ½ ton 4-wheel-drive Dodge truck that could not reach the same places a horse could go but it was more than adequate. Still, I had occasion to return to the Jackson Hole area to complete some field work and I did so at every opportunity. I was the last geologist to actively use horses for field work in the U.S. Geological Survey.

I have to admit that I still miss some of the "horseback" days. Horses can grow on you when you ride them every day and they become like a good friend. They can be lovable, adorable, and at times loyal and understanding. However, I still have memories of some pig-headed, brainless, opinionated, and self-centered horses! Every year I usually rode a different horse and they all had one common characteristic — every one was very unpredictable. One summer I had a horse get down on his knees in the middle of the trail and then try to roll over on me. Other horses have aimed vicious kicks at me while I was trying to load them. I had a scar on my chest for many years because my horse bit me.

Another geologist had an interesting experience with a horse. He was trying to load a horse onto the back of his ¾ ton truck, but the horse refused to jump on the truck bed. Finally, he placed a rope around the hindquarters of the horse to help persuade him. The horse promptly jumped onto the truck bed, then jumped on top of the truck cab and then onto the front hood on his way to freedom. Needless to say, sometimes it is difficult to explain to the front office the cost to pound out all the horseshoe dents.

The question seems to come up, "Why did you use horses in your geologic field work?" In the rugged hill country and mountains around Jackson Hole, Wyoming and the Grand Teton area, I had three choices: 1) walk, 2) use horses, or 3) find the money somewhere in order to use helicopters to ferry you around. There were only a few roads available to give access to your field areas and horses were a very efficient way of working these remote areas. First, they were cheap, costing only about $40.00 to rent per month. Second, they ate only grass which was abundant, and third, horses could get a geologist up and down a 4,000-foot ridge every day.

In 1971, after I had used horses for about 10 years, I attended a meeting in the auditorium of Building 25 at the Denver Federal Center to listen to some bright young Ph.D. from the Survey's Washington headquarters. One of the other geologists attending the meeting asked a question about using horses in field work. The bright young Ph.D.

High ridge south of Bailey Lake

proceeded to laugh and talked about how maybe that was done in the old days, but everything was now high-tech. "Horses," he said, "had no place in modern field work!" I could smile as I listened to him — just another Ph.D. out of Washington. I thought it would be nice if these guys would get out of Washington into the real world and find out what it was like — maybe he would like to climb 4,000 feet every day and walk another six to eight miles!

. . . .

As I think back over my career, I realize that I have had connections with three U.S. Presidents. The first is "Teddy" Roosevelt, the second is John F. Kennedy, and the third is Herbert Hoover; all, however, in different ways. My connection with Roosevelt is because I worked on many of his public land withdrawals. My connection with Kennedy is because I rode in a taxi parallel with him when he was driving his long Cadillac convertible in Washington, D.C. My connection with Hoover is because I once saw a dead horse; let me explain the Hoover connection.

Herbert Hoover was a young intern working on a government field project many years before he became President of the United States in 1928. He made the mistake one summer of tying up one of his mules being used as a pack animal to a very low tree branch. The mule, in trying to scratch his head with his hind foot, got all tangled up and strangled himself to death. Hoover had to pay compensation for the mule. He duly made a claim with the government controller to get his money back. His claim was denied. He tried several more times, but each time it was turned down with the comment, "No mule ever scratches his head with his hind foot." In later life, Hoover reflected on this mule incident. He said, "Whenever I see a mule, I start watching to see if it uses his hind foot to scratch his head, and you know, they almost always do."

I thought of this mule incident when my field assistant and I were riding along a trail one summer. Riding off the trail to take a look at a rock outcrop, I came across a dead horse skeleton. This horse had been tied too low and you could tell by the skeleton that he had strangled himself. My field assistant was amazed by what he saw. He kept looking at the skeleton and then walked around it several times. Finally, he said, "What do you think happened?" I said, "Apparently Herbert Hoover was here." My field assistant started to scratch his head. "You really think so?" he said.

Chapter 10

A RUSSIAN SPY

The FBI agent sorted through some of his papers from his briefcase. Looking at me, he smiled. "You probably took the elevator to the 2nd floor, right?" We faced each other across a narrow, oblong, battered old oak table that had seen its best days many years before. To his right sat a very austere, very quiet, thin-lipped older man, neatly dressed in a pin-striped dark blue suit. Another individual sat at the very end of the table. I smiled! The agent knew — and I knew — the elevators didn't run at night. Little did I know at the time that for a fleeting moment I had become a foot-note in the cold war between the United States and the Soviet Union.

The interview in the late spring of 1972 took place on the 2nd floor of one of the old red brick buildings that dotted the eastern part of the Denver Federal Center. The Federal Center had been a munitions factory during World War II and most of the original buildings had been converted over to office buildings. In the 1960s, deep grass-filled dirt bunkers still existed on the western side of the center. These bunkers at one time had sheltered bombs and other munitions. No nuclear weapons were ever known to have been housed at the Federal Center. Several 30 foot guard towers still showing evidence of machine gun placements stood as testimony to the importance of the original facility. The 12-foot high steel fence that surrounded the area during the war years has now existed into the 21st century. My office was originally in the northeast corner of Building 25, where it fronted 6th avenue.

During the 1960s, and early 1970s, access to the Federal Center had been easy, with no guards posted at any of the entrances except on weekends and nights. The National Guard and Army Reserves trained quite regularly during weekends. When the Kennedy assassination occurred in Dallas in the fall of 1963, my wife and I were living in an apartment complex north of the Center. Consequently, I went through an open part of the fence, rather than a gate, to my apartment to follow the events as they unfolded in Dallas. Frequently, I was also in the Federal Center on weekends and nights as I worked on one of my many projects. All of this was about to suddenly change.

There was one building that was very conspicuous as being completely different from the other buildings and this was Building 67. This was a very modern high-rise for its time, consisting of cement and glass about 13–14 stories high that completely dominated the western part of the Federal Center. Later buildings outside the Federal Center fence tended to mute its dominating appearance. The first floor held a bank, and a credit union occupied part of the basement. There was nothing to indicate that anything other than normal work was being done inside this building.

Rumors, however, had been circulating for some time that there was a secret government project going on in the lower level of this building. Many people tended to dismiss these rumors as having no consequences. Nobody ever commented about the heavy array of antennas on the top of the building. As events were to unfold, these antennas were probably connected through the elevator shafts to the lowest level of the building.

The FBI agent leaned forward again with much anticipation as he repeated the question. "You took the elevator to the 2nd floor, right?" I had to smile again. "No," I said, "I walked up the stairs. I wouldn't think of trying to use an elevator in a deserted building — especially at night."

Several minutes before this question, I had been handed a sheet of paper as I started to sit down. As I briefly glanced at it, I was expected to sign it essentially giving away most of my rights as an American citizen. This raised my ire considerably! I very carefully tore the sheet in half, then each piece again, and again, and again until I had a very small pile of paper. I very carefully scraped up every small piece and then leaning over the table dropped them in front of the agent. "Here," I said, "these are for you."

So how did I become — for a brief moment — public enemy No. 1? It was really my fault. I was so used to coming in on nights and weekends to work on my fossil paper in Building 25, that I had forgotten that other buildings in the complex might be different. Never once in 11 to 12 years had I ever seen a security guard in Building 25.

I had published my fossil paper and had gotten requests from all over the world asking for reprints — even from the Republic of China. This was quite an eye opener since this was in 1968, before Kissinger had gone to China at Nixon's request.

The fact that my paper had been quite successful had made me want to continue the work. The original paper had involved chip-sampling 4,000 feet of limestone at one-foot intervals and then dissolving these samples to obtain the insoluble residues. These were examined under a microscope for microfossils. All my research was done on my own time on nights and weekends.

Consequently, I had collected other samples in the late fall of 1971 and had decided to continue my investigations in 1972. It was at this point that I made my big mistake. I decided to drop off a book on mineralogy to a friend in Building 67 as I was leaving the Federal Center one evening. I walked up the stairs to the second floor in that building, but with no lights I decided it was hopeless to locate any office. As I came back down to the main floor, I saw a security guard running around and waving his arms in the air as he was shouting into his small walky-talky. Thinking that there was a major fire, I immediately ran over to him asking what had happened. He turned around and said "What are you doing here?"

Apparently, as the story developed, he had seen someone coming from the employee gift shop. I am not sure exactly what he saw but as I looked at that entrance there was no evidence of any type of breakage. He demanded to see my keys, and went through my rings of keys one by one. I said I was getting cold so we went out to my car and he watched me put my jacket on. Several days later, I received a call about somebody wanting to talk about that incident. As I went to that meeting, I was suddenly confronted by "three" FBI agents.

Obviously, they had decided they were going to blindside me — suddenly throwing accusations at me. At this, they were going to be very disappointed. When I had worked in Washington, D.C., at the U.S. National Museum as a Museum Aide in 1959, I had roomed for several months with a former cop and private detective. We had often walked at night to a small coffee shop located four to five blocks from the Pinewood Hotel where we had stayed for the summer. These walks were always fun because he would tell many stories about his previous investigations. Several of our late night discussions had also centered on interrogations, i.e., you asked questions of a "suspect" that you already knew the answers; one wanted to find out if he was telling the truth.

Even more interesting to me at that time was the peroxide-blonde that always seemed to be there with one of her clients. We usually had a bet on about whether she would have the Italian, the Latin, or the German. I usually won!

Hence, the question that was now asked, "You used the elevator, right?" Well, the elevators didn't run at night, and certainly not on a weekend. This sparring went on for about an hour. After that, I went back to my office. Within a very short time, my wife called me about three FBI agents in our house. I immediately went home and gave them 60 seconds to leave or I was calling the local police and filing harassment charges against them. They left immediately!

However, things were now starting to spiral out of control. As I discussed this odd behavior with my colleagues, their conclusion was that I needed an attorney. One of them gave me a list with names on it. He had recently been involved in court action with his teenage son who had been in an auto accident. His son had been in the back seat of a car with his girlfriend sitting on his lap. A southbound car had came across the center line around a curve and had taken off the entire left side of the car. The girlfriend had been killed instantly, and his son had received a broken neck. Beer cans had been found in the car which is why there was court action. So I picked out a name and called up the attorney. He answered his own phone and agreed to talk to me and my wife. He took the "case" and we went home quite satisfied. Within several days, I called him and he said they (the agents) had admitted that it was all "circumstantial." He said he didn't think it was going anywhere.

Several days later, I took a coffee break and opened the *Rocky Mountain News* and stared in unbelief! There was my attorney walking across the tarmac at Denver International Airport (DIA) with George McGovern. McGovern was running against Richard Millhouse Nixon for the presidency of the United States, and the attorney I had just hired was his campaign manager for the State of Colorado. As the story unfolded, the attorney had quite a background. He had served as a U.S. Attorney for 12 years, one of only 94 U.S. Attorneys that served the U.S. He had been forced out of his job after a change of administrations and had decided to try the other side of the law, i.e.., as a defense attorney.

Apparently, the hiring of this attorney really brought

out the ire of these FBI agents. My phone suddenly started giving me problems several days after I hired this attorney, with much static and some suspicious clicking that sounded like a tape recorder turning on. I remembered the advice of my detective friend who had said "If you ever think somebody is trying to tap your phone line, simply call a detective agency on that phone line and ask how much they want to sweep the line. Chances are whenever anybody hears that, they will drop the phone tap." It worked. The next day I could have called Moscow or any city in the world the phone line was so quiet.

So, why this reaction that occurred? It took several years for other information to leak out. Building 67 was rather unique at that time, situated by itself on the western side of the Federal Center. It would not have had much interference from neighboring buildings, and was also on federal property. It was the perfect place for a communication or control center. The antenna array on the roof indicated it had some role, whether as part of the U.S. warning system or just as a surveillance post. The presence of three federal agents indicates their serious security concerns about that particular building. A normal "burglary" would not require that much investigation. Apparently, federal security agencies farm out some of their activities to "satellite" offices in the United States for specific operations. Unfortunately, it seems that the ideal cover for clandestine operations are in federal buildings with many federal employees! Of the "three" FBI agents, only two showed me their identification. The third individual just sat there, never saying anything, very intent on what was being said. The CIA cannot investigate or interrogate U.S. citizens inside this country. However, they can listen in to any conversation.

. . . .

In an ironic twist of fate, years later I was summoned before two Grand Juries and the Federal Court to testify as an Expert Witness for the Federal Government in a major minerals trespass case in northwest Colorado.

AFTERWORD

At the time of the Oklahoma City bombing, I was in Elko, Nevada, helping my son build his house. As the events of that tragic day unfolded through the TV networks and newspapers, <u>I could only assume</u> that everything being reported was absolutely true. All reports indicate the suspect who was finally caught was a lone wolf, a former soldier with some psychopath leanings, who was disgruntled by the alleged U.S. Government's involvement in a deadly attack on an occult community in Waco, Texas. Was everything reported about the bombing, or were some of the facts about the bombing suppressed by the U.S. Government?

Area 51 in Nevada is an extreme example of concealment by the U.S. Government. The U2 and SR-71 Blackbird spy planes, and now the stealth fighters and bombers, remote drones, and the new Aurora spy plane were all developed at this site northwest of Las Vegas. Although this site has been reported in newspapers and TV programs, the U.S. Government has refused in the past to admit that it was a secret area. Finally in the late 1990s, the Defense Department acknowledged that many of the UFO sightings were their test planes. Lawsuits filed against the government because of environmental concerns and deaths of pilots have been historically thrown out of federal court because the government denied that area 51 existed!

In the case of the Oklahoma City bombing, the attack may have been based not only on the fact that it was a government facility, but also rumors that it was partly given over to some secret government program. Whether it did or did not doesn't really make any difference. It is the perspective of a radical militant which is the determining factor. Nothing has been written, published, or for that matter broadcast that would show a connection.

The bigger question is whether the U.S. Government would admit any culpability for the attack by having some secret security facility in this particular building? No government security agency of any kind should have any connection to a building used by civilian employees, it is just inviting terrorism plots that could kill or injure innocent people. A federal law must be passed forbidding any security concealment in federal buildings with a large civilian work force.

The recent attempt by the Justice Department to deny that certain records exist under a Freedom of Information Act request, when in fact they do exist, is just another attempt of a government agency trying to "hide its deeds." What really did bring on the bombing in Oklahoma City?

Chapter 11

AN EVENING OUT FOR THE JACKRABBITS

When you're a field geologist working in remote areas, you expect to see a lot of wildlife. In the years that I worked in northwest Wyoming, sightings of deer and elk were very common. The most memorable sighting of elk occurred in the Greys River area east of Alpine, Wyoming. My field assistant and I had ridden horseback over a high ridge and as we dropped over the other side into a big grove of aspen, the area suddenly seemed to explode with elk. The elk herd had congregated in the aspen grove expecting wind direction would provide them enough warning of danger. As we were riding horseback up wind of the elk herd, the elk couldn't decide if they were in danger or not. The elk ran past us through the trees, then turned around and ran back past us the other direction, and continued to mill around. Finally the main herd thundered down the ridge away from us.

The most phenomenal wildlife displays of small game I ever witnessed, however, came in the late 1970s, when I had moved to the Evanston, Wyoming area to work on coal deposits. I had taken charge of several drill rig crews that were drilling out federal coal leases for evaluation. At one location, I had a crew drill a 200-foot-deep hole in a small pasture to evaluate a coal seam that was not exposed. The rancher, who had given me permission to drill, had complained about his beef cattle not have much water that year. Water, however, had come up the drill hole almost to the surface and would provide an ample supply for his cattle that summer. I offered the rancher the use of that drill hole if he wanted to put down 4-inch plastic pipe and cement it in place to make it his water well. He very eagerly accepted my proposal.

A week or two later I was returning to Evanston from Kemmerer, Wyoming, after darkness had fallen. I decided I should check and see if the rancher had put in the plastic pipe for his water well. As I drove off the main highway on a small dirt road, I came over a small sandstone ridge into the pasture area. I immediately stopped the truck as I couldn't believe what I was seeing. The headlights of the truck showed not just dozens of rabbits in the pasture area, but what literally seemed like hundreds of rabbits. There were so many rabbits that they resembled a herd of sheep. As I finally started to edge into the pasture area towards the well site, the rabbits seemed to explode right and left of the truck like a flock of birds. Some of them seemed to jump as high as the hood of the truck. This scene continued all the way across the pasture area to the drill hole location.

After reaching the drill site, I determined that the rancher had put down his plastic pipe and had cemented it in place. Getting back in the truck I expected to see the same number of rabbits as I left the pasture, but only three to four were visible. I had been stunned to see the number of rabbits when I drove into the pasture area, and I was very surprised to see them disappear. The whole experience seemed unreal!

Later that week I couldn't contain myself and I had to go back to the same pasture area and see how many rabbits I could find. I walked up and down the gullies and small valleys; I walked over the ridges; I scoured all the hillsides. It was hard to believe, but I did not see a single rabbit, nor did I locate any of their dens. Where they had come from and where they suddenly disappeared to has to be one of life's greatest mysteries. Do rabbits ever get together for a convention?

Chapter 12

POLITICS WIN — SCIENCE LOSES

$20 Million Coal-Lode Maps Useless

On April 29, 1979, I walked out the front door of my house on a beautiful spring morning to pick up the *Denver Post* newspaper. The headlines screamed at me: "$20 Million Coal-Lode Maps Useless." The article went on to quote critics who said that poor planning, bad management and inaccurate work resulted in a multimillion dollar set of unreliable and useless maps.

I was stunned to see the headlines, but not really surprised. There had been much dissension among the geologists within the Conservation Division for the last several years about this new program. The program had been designed to spend $20 million to produce maps showing the occurrence and development potential of coal resources on 1,400 "quadrangles" containing federal coal deposits. The program proved to be a complete disaster from the very start.

The *Denver Post* article was factually correct, but wrong

in one very large regard. The program was really "politically" driven, not science driven by the U.S. Geological Survey as the article suggested. The *Denver Post* article mentions the U.S. Geological Survey several times as running the program, as the Conservation Division was technically still in the U.S. Geological Survey. The Conservation Division, however, had basically been taken over by "political appointees" made by the Carter administration who had no knowledge of minerals or mapping techniques. The immediate chief of the Conservation Division was a "political science professor" from Oklahoma. The political action of placing individuals with no knowledge in the mineral resource field resulted in poor planning and bad management decisions which destroyed what could have been a valuable and useful product.

I need to explain how this type of government program started, and also how the politically driven concepts concerning the leasable mineral programs on federal lands led to the demise of what could have been a valuable asset. It is also of some interest to consider how part of the letter I sent to the Director of the U.S. Geological Survey concerning this program got quoted in this front page article without my permission.

When I moved to southwest Wyoming in the middle 1970s, I knew that there was going to be renewed emphasis on the coal program in response to the Arab oil embargo. Until this point, I had not been affected by the political events sweeping the country. The turmoil over the Vietnam War and Nixon's resignation were not events that would affect the programs of the Conservation Division. With the election of the Carter administration, coal suddenly became more significant to the nation's economic outlook. The political winds that were to follow would have significant impact on the future of land-use planning involving our national mineral resources.

My move to Evanston to work on coal resources in the surrounding area had been relatively easy for me compared to previous years. My family lived in the same two trailers I had used in the Greys River area. Now, however, there was no need for horses, saddles, and other riding gear or camping equipment. I didn't have to plan where I was going to corral my horses or put my horse trailer on the weekends. All I had to worry about was keeping gasoline in my 4-wheel-drive Dodge truck; of course, there was always my field assistant . . . The trailer park in Evanston was in the eastern edge of town. It had nice wide spaces where I could park my trailers, as well as mature trees that provided ample shade during the hot summers. The town had a nice swimming pool for the children. My oldest daughter had a part-time job in the park motel which helped to keep her occupied. There was an old field in the park where I could play football with my son. Salt Lake City could be easily reached for the many amenities which it offered.

Field work in this area was much easier than before; gone were the 10 to 12 hour days on horseback. The terrain could be described as more "hill like" instead of the high rugged peaks around Jackson Hole and the Tetons. The geology was much easier to map compared to where I had been working. I could reach most areas with my 4-wheel-drive Dodge truck. When I had to stop my truck because of no road access or locked gates, it was an easy walk to an outcrop or to measure coal beds. But the old sage was right,

"And this too shall pass away!"

The focus of the minerals program suddenly changed. A directive came down from Washington to evaluate Federal Coal Leases in all areas of the West to determine their mineral value. I developed a drilling program for southern Wyoming to drill and evaluate coal leases from the Evanston/Kemmerer coal field, through the Rock Springs area to the Hanna Basin coal field near Medicine Bow, Wyoming. I had charge of two drilling crews for parts of two summers. My family stayed in Denver the second summer while I was constantly hauling my trailer from location to location. Depending on the location, drilling activity might be 10 days straight with no time off. I managed to get 61 holes drilled, but I could have doubled that number if there were not constant breakdowns with the equipment. Drill holes would average between 200 and 1,000 feet in depth, with electrical logs run on each location for coal evaluation. With the old and decrepit drilling rigs, I was also concerned that none of the drilling crews get seriously injured.

As the drilling of coal leases progressed, the Washington office wanted maps prepared that showed the Coal Resource Occurrence (CRO) and Coal Resource Development (CDP) for all areas. The project was planned to help the Bureau of Land Management (BLM) in its land-use planning procedures for the leasing of federal lands for future coal mining. Unfortunately, manpower was not available in-house for the job. Contractors were now hurriedly hired to prepare these maps. The problems would now start to escalate. Complaints would start to pour in about the quality of the maps.

The main problems identified on these maps were the following. However, there were many more:

1. **Contracts given on areas with no coal resources** — The geologists in the Conservation Division were always puzzled about contracts being issued for some areas which had no coal resources. A contractor would receive $10,000 for producing a blank map that covered 54 square miles, or approximately 35,000 acres. Concerns were raised that this may have been payoffs for political contributions, but this idea was quickly discarded. It was also difficult to phantom why contracts were also given on quadrangles which contained only one data point, usually an exploratory drill hole that a coal company may have drilled in the past. Again, the contractor got $10,000 for basically issuing a blank map. Like the fabled "Bridge to Nowhere" in Alaska, it is difficult to determine the exact cause of why something is done.

   The CRO and CDP maps were politically driven and a good example of what will happen when scientific analyses are ignored. The political appointees did not have a sufficient scientific background to understand what was needed for the program to be successful. No project can be managed successfully without placing knowledgeable individuals in charge of the program. A competent manager would have immediately made changes in the program, but with no to little background in mineral management, he was unable to respond as he had no concept of what was needed to resolve the problems as they occurred.

2. **Inaccurate maps** — The major problems resolved around inaccurate maps. Maps were compiled from all sources, even old data that was never verified. Coal beds and potential mining areas frequently did not match from map area to map area. A coal bed would extend up to a map boundary, and suddenly not appear on the adjacent map. Boundaries of potential mining areas did not always match. The contractors complained that they were not allowed to field check maps as the contracts were written to use only existing data. Speed was important, accuracy was not. Inaccurate maps are exactly that, inaccurate maps. They cannot be used in land-use planning decisions because the data is not known to be accurate or not. The geologists in the Conservation Division tried to correct many of the mistakes made on the maps but finally gave it up as there were too many corrections.

3. **Maps didn't show private or state-owned coal lands** — When planning coal mining operations, a coal company would need to know how the federal leases would fit into the total mining plan. The value and importance of these maps were greatly diminished without this information. It is equivalent to buying a road map that shows only county roads, but no state or federal highways. One doesn't know how to travel without more information. Boundaries of adjacent coal land are extremely important as well as who owns or controls those leases.

4. **Maps already existed that gave the same data** — A contractor would continue to luck

out. He would get paid $10,000 for basically copying another map. When Ronald Reagan was running for President, he cited his definition of a "Bureaucrat." "It's a person," he said, "who notices a road sign that is either too high or too low. He immediately tries to figure out whether he has to dig up the road to match the sign or bring in more material to raise the road."

5. **Some maps contained information of little value in land-use planning** — Ronald Reagan again.

6. **Some coal beds had wrong legal locations** — Inaccurate information is always a curse to the user. Nothing can be worse when making plans for a drilling program than to suddenly learn that coal doesn't exist in that area you may have just leased for a future mining operation. In this instance, the contractor would have to bear the blame for this error. Whether coal-bearing formations exist or not in an area should be of general knowledge to anybody with a geologic or mining background. However, the Interior Department must also bear the responsibility for putting out inaccurate or misleading information.

7. **BLM offices in some areas already had the information they needed** — The fact that many BLM offices already had the information they needed for land-use planning is indicative of the poor planning and lack of knowledge of the political appointees anointed by the Carter administration to run the program. It is almost impossible to believe

that a program of this magnitude, eventually costing over 100 million dollars and serving as a model for other leasable minerals, would have had no definitive planning during its inception. There was no concern voiced about trying out a smaller pilot program to work out any potential problems. The CRO/CDP maps now sit in an isolated warehouse gathering dust, a monument to government bungling, inefficiently, and political foolishness.

8. **Coal companies complained about the cost of the maps** — Coal companies found they could get the same data, as well as more accurate data, for a lot less money by doing some work themselves. The question has never been answered, "Why are maps produced by the Interior Department that are unusable, and then cost more than what a private company would pay for them <u>even</u> if they could use them?" Of course, the answer is politics!

The newspaper publicity finally resulted in the entire program being scrapped. Apparently, the 20 million dollars allocated for this type of program was only supposed to be the beginning. I had opposed this program from the beginning and had written a letter to the Director of the U.S. Geological Survey expressing my disgust about the quality of the maps. My concern had developed when I had visited a contractor's office where the CRO and CDP maps were being produced. The room was about 10 feet by 12 feet, with six to seven desks pushed together with a number of college students standing shoulder to shoulder compiling these maps. The scene indicated that no quality control was being exerted by anybody. This type of environment would not happen under competent management. Unfortunately, the political process of interfering in land-use planning by political appointees would extend into the 21st century.

In retrospect, it would seem obvious that the internal turmoil caused by the map controversy would have been sufficient cause to cancel the program before it hit the headlines of the *Denver Post*. Even though there had been much dissention among the geologists concerning the poor quality and inaccurate data of the maps, the political appointees were unwilling to cancel the contracts. I was at the meeting where the official made the statement, "We may have to settle for something better than crap." At that same meeting, he also indicated that this program would continue, as he didn't have enough guts to throw away ten million dollars.

The fact that the official had already thrown away the ten million dollars didn't seem to concern him, only that his pride might be hurt. In today's world, the ten million dollars would probably be closer to about thirty million dollars. It was his total disregard for scientific accuracy and the taxpayer's money that resulted in my famous letter to the Director of the U.S. Geological Survey. In part of that letter I had stated that, "When various companies spend thousands of dollars for these maps and find out they are for most purposes unusable, there will be many repercussions." Other geologists turned this letter over to the reporter, and it was used without my permission.

The unfortunate part of the whole controversy was that it affected the scientific investigations of the U.S. Geological Survey by reducing its funding requests that drastically cut its programs. The U.S. Geological Survey was not responsible for the fiscal irresponsibilities and bad decision making by political appointees of the Carter administration. It's name got linked to the scandal as somebody had to bear the blame, not the politicians. This inattention to scientific accuracy by another political organization, the federal Bureau of Land Management, will eventually lead to a vast environmental disaster in northwest Colorado.

Chapter 13

SLEEPING WITH THE ENEMY

We don't need scientists — All we need are
common, ordinary, run-of-the mill bureaucrats

In the late 1970s, it was rumored that a Federal Bureau of Land Management (BLM) official had said, "We don't need scientists; all we need are common, ordinary, run of the mill bureaucrats." This comment, whether true or not, helped to cement the gloom and despair in the earth science community within the federal government. Not too long before, a low level Minerals Management employee on the Gulf Coast had been elevated to the position of Assistant Secretary of Energy and Minerals. It again was rumored that this was in response to a donor in California giving five million dollars to the Jimmy Carter election campaign.

This personnel action then led to many problems in the understanding of how federal lands should be managed. When the Assistant Secretary made public statements such as, "If you want more oil and gas, just drill a bigger hole in the ground," it immediately brought industry indignation. The Assistant Secretary was booed and jeered off the stage at various times when trying to give talks on energy policy. Memorandums came across my desk from the Department of the Interior that forbade the mention, discussion, or any memorandum in which the name of the Assistant Secretary might be used!

The Bureau of Land Management has a mandate, not only from the U.S. Congress but also the American taxpayer, to protect the federal lands from unsafe use and exploitation by making in-depth scientific studies. When proposals are made by private interests to develop federal lands for their mineral values and this procedure isn't followed, tragedy can only result. In the following pages I will document an actual case of noncompliance in northwest Colorado to illustrate this problem. But first it is necessary to examine how the BLM received this mandate for land-use planning for mineral development on federal lands.

Until the late 1970s the BLM could be regarded more as a horse and cow agency than anything else. The agency issued permits for cattle grazing on federal lands as well as keeping track of federal and private mineral rights. From time to time it also published maps which showed the land status of federal and state lands and made the maps readily available to the public for a small fee.

In the late 1970s a change occurred in a radical governmental reorganization that altered land-use policy. Decision making, based on political policy that became the main priority, was to be made by the BLM, with only "lip service" given to scientific evaluations. When one particular political party holds power, their policies will translate over to decisions made by the BLM concerning the public lands. Land-use decisions become a "political football." Decisions as to "Do we permit leasing in this area?" or "Do we permit development?" are not subject to scientific evaluation or extensive scrutiny, and thus lead to potential environmental disasters.

In any land-use decisions made by the BLM, it will be sure to point out that, "We always hold open meetings, where we obtain public comment on any action," to justify its final decision makings. What they don't want known is that in any of their "open meetings," about the only individuals turning up are usually local environmentalists complaining about how a bald eagle nest will be impacted. Occasionally a local rancher will be concerned about road access to his property. In other words, no scientist turns up to dispute the proposed action, so no adverse studies are ever on record. This allows the BLM to claim that, "We have studied all options."

It goes almost without saying that almost everybody is familiar with the past environmental failures. The asbestos scare, for example, has caused major health concerns. The failure of the government to recognize the danger from development of nuclear weapons and private development of nuclear power gave rise to the SUPERFUND, where billions of taxpayer dollars were spent on the cleanup of waste sites. Mines that were opened for the extraction of uranium ore, as well as others being mined for gold, silver, copper, etc., have in many cases continued to leach dangerous elements into our water sources.

During this time of upheaval in the late 1970s I was working in the Conservation Division of the U.S. Geological

Survey. This group was mainly involved in the classification and scientific evaluation of leasable minerals on federal lands, such as oil and gas, coal, phosphate, sodium, and other lesser mineral. The responsibility of that Division were briefly transferred to Minerals Management and then to the BLM. Unfortunately, the BLM was not prepared to assume any mineral responsibilities under political mandate.

Under the merger scientific professionals were gradually disposed of in various manners. The former Regional Geologist of the Conservation Division, George Horn, was given a stripped down barren office with only a chair, desk, and with no pencils or telephone, so he soon retired. Other geologists quit, while others found jobs in other government agencies. In my case I escaped to the oil industry to avoid the turmoil.

In 1985 I returned to government service after the collapse of the oil industry and in the process joined the old Conservation Division office in Tulsa, Oklahoma, only now it was called BLM. The only noticeable change was the sign on the office door. Science still ruled the day. However, the lure of the mountains in Colorado was too much to ignore and I obtained a transfer to Meeker, Colorado, in December of 1988.

When I first arrived in the town of Meeker, I was immediately impressed with the surrounding countryside. The area is one of pristine beauty and solitude, with one expecting to see million dollar homes dotting the area, just like the Jackson Hole country near the Grand Tetons of northwest Wyoming. The White River flows just south of Meeker and extends eastward over 40 miles. Dude ranches occur along this section of the river, with fly fishing in the White River one of the major activities for tourists. Westward the river flows past Rangely, near the Utah boarder. Ranching is the major source of economic support. The entire area is inundated by hunters for elk and deer during the fall hunting season. Major development of the area has been hindered by the oil shale "boom and bust" cycle of the 1970s and 1980s.

The town of Meeker, however, has had a major problem since its existence with the cracking and sliding of many building foundations. Many of the houses in town have become unlivable in the past and had to be torn down. The problem lies in the fact that the entire town area is underlain by a thick section of black, impervious shale rock. A thin veneer of hill wash and sediment of uneven thickness covers this shale unit. Water flowing downward from the western hillside on top of this shale and underneath the soil cover has frequently caused movement that will affect building foundations. The only way to stabilize the foundation of buildings is to pound steel or cement casings deep into the shale rock for support. The expense of an additional $20,000 for building a house or other buildings becomes prohibitive for most families, especially with a limited economy in a rural area. Development of mineral resources are seen by local residents as a desirable way to create jobs and pump more money into the local economy.

The structural damage to housing in Meeker however, is indicative of the major problems that are going to develop in the surrounding area due to a lack of adequate engineering studies involving the subsurface development of mineral resources. In addition, a lack of understanding of the geologic framework of the area, as well as the geologic process involved, compound the problem.

All the environmental failures mentioned so far, however, have one common thread running through them. They are all repairable when enough taxpayer money is thrown at them! But what happens when hazardous conditions are created by private development of subsurface minerals where geologic conditions are set up that will cause devastating problems in future years? — geologic conditions, for example, that will continue to exist no matter how much money is thrown at this problem. It is this scenario that will play out in northwest Colorado because of inadequate management and supervision by BLM. So, how does an environmental disaster start?

. . . .

The male voice came loud and clear over the phone. "I sent a letter over to BLM in Meeker over a week ago opposing the subsurface mining of sodium minerals unless additional engineering studies are done. Didn't anybody read my letter?" This phone conversation with a former structural engineer who had worked for the U.S. Bureau of Mines came several days after I had returned to the office after completing some field work. I had noticed a big jovial office meeting with consultants going on in the conference room and had asked the secretary what was going on. "Oh," she replied, "the office is giving the final OK for the subsurface mining of sodium leases." I stood there in an absolute stunned silence. Nobody in the office from the manager on down had ever discussed or brought up the subject of sodium mining with me during the time that I had worked as a geologist in that office.

My main job in the Meeker office up to that point had been the determination of oil shale horizons in the Piceance Creek basin west of Meeker. Various different levels needed to be cemented off during oil and gas operations to prevent any impact on the oil shale. Any subsurface mining of sodium minerals could vastly affect the future of oil shale development. While development had failed recently, new methods of oil and gas extraction from oil shale could possibly be developed. According to the Rocky Mountain Association of Geologist guidebook, 1974, p. 188

> **In 1968, Secretary of the Interior Udall signed an order placing limitations on the leasing of oil shale lands for sodium minerals by requiring that such permits or leases by issued only where the Secretary of his delegates find that it will not adversely affect the oil shale values of the lands and only for sodium beds which can be worked without removal of significant amounts of organic matter.**

Although sodium leases were later issued, it was obvious that there was much concern about mining activity. Under Secretary Udall's 1968 order, the BLM has failed to follow this requirement that oil shale lands be protected!

Oil shale and sodium minerals were deposited together in an old structural area that was once part of an ancient fresh water lake called Lake Uinta. With a climate change that old lake became very saline and started drying up, leading to the deposition of oil shale and sodium minerals. The oil shale has been estimated to contain 1,200 million barrels of oil-equivalent when fully developed. Sodium minerals, nahcolite and dawsonite, are found in the very depositional center of the old lake. Nahcolite can be used in many industries, but the public knows it in the grocery store as baking soda. Dawsonite is slightly different in its makeup, containing an aluminum ion.

Nahcolite reserves are estimated to be 29 billion tons, and Dawsonite reserves at 19 billion tons. Halite (rock salt) reserves are estimated to be in the 25 billion ton range. The two oil shale tracts, Ca and Cb, received bonus bids of $329,083,600 in 1974 when they were put up for lease.

Since I had been completely ignored in any decision making, I tried to find out more information about the mining process. First of all, nobody knew anything about a letter from a structural engineer. More amazing, nobody seemed to know anything, or were willing to discuss anything, about which they had just agreed to! I found this lack of knowledge on everybody's part to be absolutely fascinating.

There was no doubt that their apprehension of my being involved was well founded. My name was on geologic maps that covered almost 1300 square miles of some of the most complicated geology in North America. I had spent four years in the Tulsa, Oklahoma, office involved in making subsurface maps showing oil and gas potential of parts of east Texas. I had also worked in the petroleum industry for several years making structural models of the northern part of the Idaho-Wyoming Overthrust Belt using all available data including seismic interpretations. Now in the latter part of my career, I was working the Piceance Creek basin to protect oil shale reserves where the sodium mining was going to take place!

Without any information given to me about the mining project, I was forced into spending my time digging through the library files to obtain data. Reviewing geologic reports was something I had done on a regular basis while working in the U.S. Geological Survey. All of my maps and reports had gone through a rigorous review process before publication, and I had also acted as a reviewer on maps and reports made by other geologists. The job of any reviewer is to make sure that there are no "loose ends" in that report or map. Not only was the reputation of the U.S. Geological Survey at stake, but also that of the individual geologist as well.

The mining plan and supporting data indicated one major problem — WATER had not been adequately addressed in the consulting reports. Only one drill hole had been used, that to show water movement downward — but other in-house reports as well as published data showed water movement upward, particularly during wet cycles. When I pointed this out to the Minerals Supervisor, his response was simply, "Well, it's too late to do anything about that now." Apparently he was not even aware of this data!

Water movement in the Piceance Creek basin is extremely important. The removal of too much rock material from the subsurface may cause overlying rock formations to start fracturing due to slumping caused by lack of structural support. Subsurface water that was previously confined would now escape to the surface. While salt deposits were laid down at the bottom or depocenter of this old lake bed, it would now become mobilized! With fracturing and faulting of overlying rock beds, salt waters would start pouring out on the surface. Once the fracturing and faulting of overlying rock units had taken place, there is no way to shut off the upward movement of salt brines to the surface. All of the salt eventually would wind up on the surface and into the White River and the Green River — ALL 25 BILLION TONS of it!

The Drilling and Development Plan for development of sodium minerals had been very simplistic. Multiple drill holes would be drilled in a set pattern, so many feet apart. When reaching a certain horizon, they would be drilled in a horizontal plane, in an interval of nahcolite-bearing saline beds. After a certain distance, these horizontal drill holes would be tapped by vertical wells. Water would be pumped through these horizontal drill holes under pressure dissolving sodium minerals. Eventually all fluids would be pumped out at the second location and run through a dryer plant. The distance between the original surface wells was to be maintained throughout the mining process. This was to ensure sufficient structural support throughout the mining process.

Unfortunately, the mining plan had a very strategic flaw — the saline minerals didn't read the mining plan! Sodium minerals, as well as salt, are very soluble and will dissolve very easily. There can be no firm predictability as to where the saline minerals will form the necessary support to prevent fracturing and slumping of overlying rock units.

At the time I retired in early 1997, I finally succeeded in reading one of the mining reports. It reported that cameras had been sent down through some of the horizontal pipes that showed water in the various drill holes was starting to communicate with each other. The structural support that was supposed to be there was now being reduced to a minimum. Structural support was now occurring in a haphazard manner, leading to an inherent weakness throughout the entire mined area.

When I left Meeker in 1997, I was aware that oil and gas development would soon start in the Piceance Creek basin west of Meeker. My prediction became true and that area has now become a favorite target of the oil and gas industry, which is using extensive hydraulic fracturing for the oil and gas production. In this process, millions of gallons of chemicals and water are pumped under pressure into oil and gas wells to fracture rock units and permit hydrocarbons to be produced. Unfortunately, the chemicals used in "fracking" are unregulated and can be dangerous when introduced into the drinking water of local residents. Currently, the BLM does not regulate the chemicals being used in "frac" jobs, allowing industry to use anything it wants to. Even if regulations are eventually introduced to stop dangerous chemicals from being used, it will be too late to stop the contamination that has already been introduced into the water system.

Before I left Meeker in 1997, I climbed a hill overlooking the town and surrounding area and only stopped when I needed to drink in its pristine beauty. My thoughts were on when the desolation would start; would it take 50 years, or 75 years, or more? As I wondered about it I knew one thing for certain; some day it would start — a comic opera, as the poet would say, of gross mismanagement by the Federal Government!

The desolation would start slowly at first. The ranchers would start noticing an increase in salinity in local streams and ground water. They would also start noticing a strange oily type smell to drinking water and start hauling in some drinking water for their use. Cattle would start losing weight. Natural gas would start bubbling up along hillsides where there were some white-coated salt fractures in the rocky hillsides. Children would be sent to live with relatives because they started breaking out with scabbing and lesions all over their bodies. Cancer concerns would grow as the general health of the local ranchers started to deteriorate at an alarming rate. It wasn't just the salty water that was causing the problem; there seemed to be something else, some odd tasting chemical!

As the contamination process speeds up, water will become completely undrinkable, and vegetation will suffer. No cattle will be able to graze in the area. Hunting will suffer, and chemical pollution of the water system will mandate no elk or deer meat can be eaten that come from the Piceance Creek basin area. Tourism will suffer, and the local economy will take a serious downturn. All of these problems will be magnified by future coal-bed methane development in the areas around the town of Meeker itself.

Still standing there on that hillside many years ago, I was reminded of a local resident who had commented to me about wanting more industry to come to that area. "We need jobs," he said. "Nothing bad can happen here. We want to have it all." Indeed, it now seems he will get his wish!

. . . .

The "worst-case" scenario presented in my story is a real

possibility unless additional steps are taken by BLM officials to prevent a major environmental disaster. <u>Up to this point, no known damage has occurred or has been released by the BLM officials!</u> The question will always remain, "How much monitoring does BLM need to do of the mining process?" Before I retired, I had to dig through mining reports to determine that the water in various drill holes was starting to communicate between all the drill holes — a serious breach in the mining process. Yet the BLM office staff didn't show any concern!

All oil and gas companies who have a vested interest in oil shale in the Piceance basin must be notified and kept informed of all mining plans concerning the extraction of sodium minerals. The oil shale in the basin is estimated to contain 1,200 million barrels of oil equivalent. If oil is $100 per barrel, the oil shale in the basin is worth well over <u>one trillion dollars</u>. If the mining of sodium minerals for a baking soda product prevent oil shale from being developed, the country could suffer a great loss.

A detailed structural analysis needs to be conducted to determine the tipping point, i.e., a point where too much removal of subsurface material will result in the slumping and fracturing of surface rocks. Other monitoring procedure include the drilling of many water wells to check on upward movement of subsurface water as well as their salinity levels, particularly during "wet cycles." Oil and gas companies must be required to list the toxic chemicals used in their "fracking" procedure for oil and gas production, as well as the <u>total volume</u> introduced into the subsurface. Monitoring activities must be conducted by a consulting group that is accomplished in this type of work.

Procedures must be set up to show how leakage of brine-filled waters and/or toxic chemicals would be confined if a leakage occurred. Finally, all information about sodium mining must be made available to the Colorado Geological Survey. The state of Colorado has a vested interest in the mining of sodium minerals even though the mining occurs on federal land.

The average oil and gas well that requires "fracking" uses 5,000 gallons of chemicals in the process of hydrocarbon

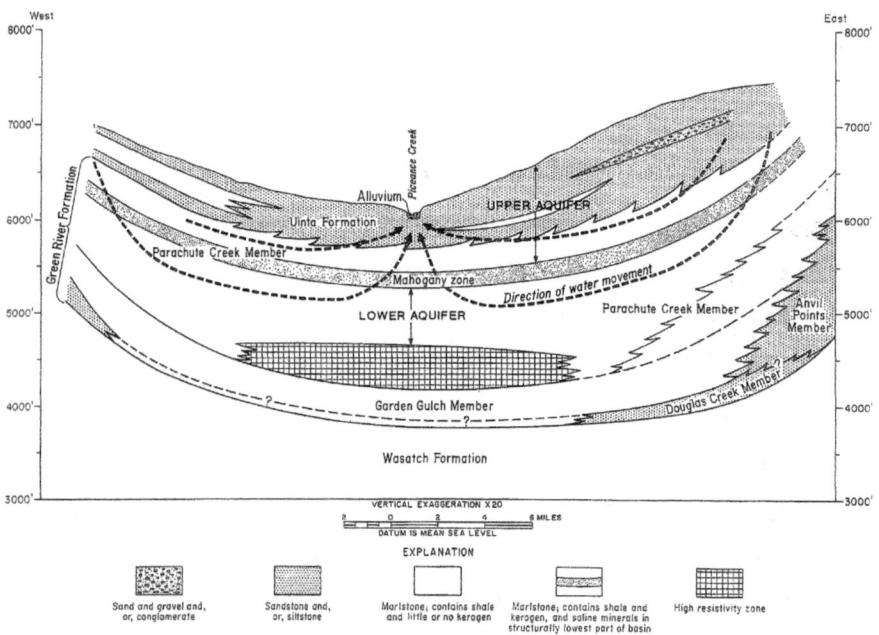

Diagram taken from "Water Resources of Piceance Creek Basin" by John Weeks, Energy Resources of the Piceance Creek Basin, Colorado, p. 177, Twenty-fifth Field Conference Guidebook, 1974, Rocky Mountain Association of Geologists.

production. Of the 200 different types of chemicals that may sometimes be used, 90% are known to be dangerous.

The high resistivity zone (shown cross-hatched in the

above diagram) consists of sodium minerals, halite, and oil shale. The fracturing of rock units will cause mobilization of halite (salt) into the Upper Aquifer and into surface rocks and streams. Halite is estimated to be around 25 billion tons which will ultimately be dumped into the White and Green River drainage systems. In addition, the ultimate development of oil shale may be compromised because of the mining of sodium minerals. An estimated 1,200 million barrels of oil-equivalent are in the oil shale deposits of the Piceance Creek basin (National Petroleum Council, 1973).

AFTERWORD

It has become readily apparent that the feeling, "We don't need scientists; all we need are common ordinary, run of the mill bureaucrats" can no longer be tolerated. The governmental process can only proceed in a responsible manner when policy is put in place that is realistic. There are certain obligations that must be made to the American public. In this regard then, it is quite obvious that the BLM cannot function in land-use planning when science is ignored. It simply does not have the capability of adequate geologic staff with sufficient scientific background to analyze the problems that exist in land-use planning and that will continue to come up in the future. The only answer is to place anything dealing with minerals and mineral development back in the hands of the U.S. Geological Survey from where it originally came.

The disregard for research is again shown in 2012 by BLM when it approves new leases for oil shale development without doing sufficient studies to determine the feasibility of their proposals. One company, for example, would now hydraulically fracture oil shale horizons to theoretically enhance oil recovery. The fracturing of additional rock layers will now permit easier migration of brime-filled fluids carrying toxic chemicals to the surface besides that produced by sodium mining. The diagram of the "Water Resources of the Piceance Creek Basin" by Weeks illustrates the relationship of salt deposits to the rest of the basin.

Recent newspaper articles show developments concerning sodium mining.

1. On September 7, 2011, the *Denver Post* carried an article about a company that wanted to double its production of nahcolite in the Piceance Creek basin to 250,000 tons per year. The company noted that it plans to build a processing plant at a cost of 34 million dollars near the town of Rifle to even expand that production level.

2. A mining engineer has estimated that one ton of saline rock is approximately one square yard. If the 250,000 tons of nahcolite is to be obtained, the square yardage would have to be doubled, because of the interbedding nature of nahcolite with other rock, giving 500,000 square yards of mined material. This would translate into an area of 30 feet by 675 x 675 feet. Ten years of mining (suggested by the 34-million-dollar processing plant) would produce a much bigger mined area underlying the oil shale horizons. This is in violation of Secretary Udall's order and would possibly destroy the future mining of oil shale.

3. The Colorado Geological Survey has stated in an article that it is now investigating sink-holes and rock fractures and other indications of slumping over abandoned coal mines in southwest Weld County in Colorado.

CHAPTER 14

A TOWN THAT COMMITTED SUICIDE

*The town of Centralia lighted a match —
and the town crumbled*

The ANTHRACITE MUSEUM in Scranton, Pennsylvania, tells a very disturbing story about a town that destroyed itself. This story is disturbing for two reasons. First, it indicates the tragic circumstances that can result when the geology of an area is ignored. Second, it demonstrates that the devastation that occurred in and around the town on Centralia, Pennsylvania, can also occur in the area west of Meeker, Colorado, this time because of political indifference to scientific research. A daunting geological parallel can be drawn even though these two areas are in different parts of the country.

Centralia was a typical town in the coal country of northeast Pennsylvania. Numerous subsurface coal seams had been mined and abandoned in the past as the subsurface mining progressed throughout the area. The town itself sat over many of the abandoned mine workings. The story of this tragedy starts in the early 1960s when a young teenage boy noticed steam coming from the ground around where he lived. Drawn by what he saw, he tried to determine the cause. As he got closer to the steam area, the ground started to cave in around him and drag him into the abyss. His frantic cries brought help from his cousin which prevented him from being dragged 300 feet to his death.

As an investigation into the tragedy started to unfold, the story of the young boy almost dragged to his death was minor compared to the problems that started to occur throughout the whole area. The investigation showed that in past years the town authorities had given permission to have refuse and garbage thrown into a sinkhole near the town boundary. Finally, the amount of refuse in the depression grew so big that a decision was reached to burn everything in the pit area. The fire department was given the assignment to extinguish the fire after the burning of the refuse was near completion. The burning of refuse however, started a chain of events which led to the destruction of the town.

The town authorities <u>did not know</u> at that time that the sink hole was connected to some of the underground mine workings that still contained a considerable amount of coal. The burning of refuse ignited fires that spread throughout the old mine tunnels. There was no way to extinguish the fires even though many efforts were tried. The fires burned many of the mine tunnel supports which were composed of unmined coal, as well as timber supports, thus leading to the slumping of the overlying rock formations.

This slumping of surface rocks now caused rock fracturing which allowed various coal gases to escape to the surface and affect the health of the town residents. The health problems increased to the point where the federal government finally intervened in the early 1980s to provide 42 million dollars of taxpayer money to buy and remove the remainder of the 600 homes in the town area. The destruction of the town was now complete. The coal in the old mine workings is still burning and this process will continue for many more decades. Photographs of the area resemble a World War II battle zone.

The geologic connection between Centralia in northeast Pennsylvania and northwest Colorado may seem somewhat remote at first glance. Centralia lies within a coal mining area where traditional mining methods were used. In northwest Colorado, sodium minerals are currently being extracted by using solution mining methods. Different minerals and different mining methods would seem to imply that there is no similarity between these two areas. What is similar in both areas, however, is that both mining methods will allow slumping and then fracturing of surface rocks due to the lack of structural support at depth. In northeast Pennsylvania, the toxic agents causing the health problems were methane, carbon monoxide and other coal gases seeping through fractured surface rocks. When the environmental disaster starts in northwest Colorado, it will be salt-filled waters carrying toxic chemicals used in the "fracking" process by the oil and gas industry also seeping through fractured surface rocks.

Under normal circumstances, the "fracking" of shale intervals in the Piceance basin would not cause any problems as this activity would be conducted at a depth where all chemical agents would be confined. Due to the variability of sedimentary rock intervals throughout the basin area as shown by cross-sections in geologic guidebooks, it is impossible to predict how the oil and gas development will affect the basin environment. However, there can be no argument about the fact that the strong eastward flow of subsurface water in the basin from the western highlands will increase the movement of all elements into fractured sedimentary intervals in the central and eastern parts of the basin, especially during wet cycles.

The problem with "fracking" is that the initial production rate is quite high, but will rapidly fall off, making it necessary to drill many production wells. If an average of 8 to 12 horizontal bore holes are drilled from one production platform, over 60,000 gallons of toxic chemicals are dumped underground in a fairly small area. In an average field, with one production platform every 160 acres, there would be 40 horizontal "fracking" wells, using a total of 200,000 gallons of toxic chemicals per square mile.

"Fracking" for natural gas production from sub-surface shale intervals, may occur up to 10 times to keep production at a higher level. In the example cited, up to 2,000,000 gallons of toxic chemicals could possibly be used per square mile. Although excess fluids are pumped into holding ponds, evaporation will leave toxic chemicals exposed to wind action that will contaminate surrounding areas. Burying toxic sediments in place only delays their release into surrounding drainages. A 10,000-well field could use up to 500,000,000 gallons of toxic chemicals — causing extensive pollution to rock intervals containing fresh water.

In any area where significant drilling is to occur, a study must first be done to establish a baseline of water quality throughout all fresh-water zones. This is necessary to determine the extent of any damage to those zones if pollution problems occur in the future.

CHAPTER 15

THE "PHANTOM" WILD HORSE

Why angels want to weep

The photo of a wild horse hanging on the wall of a Federal Bureau of Land Management (BLM) office in northwest Colorado was one of the worst I had ever seen. The poor horse looked gaunt, obviously underfed with his ribs sticking out of his sides, and other scabs and lesions very conspicuous. Worst of all, his mane and tail looked like they had been caught in the door of a commuter train. The BLM Ranch Tech caught me looking at the photo as he wheeled around in his office chair. "It's time to go out there and round up those little darlings again," he said. Looking at him very closely, I said, "Are you sure that horse doesn't have parasites?" Placing his hands in back of his head as he leaned back in his chair, he said, "Yup, probably does." I said, "Do you think you can find any wild horses to round up? He said, "Well, if there are any out there, we'll find them."

At that point I had to suppress a smile. Several days earlier, one of my office mates had come to me and whispered that, "One of those guys switched the tailgate on his truck with yours. He damaged his tailgate by backing into a steel gate. Since both of you have identical trucks, he switched tailgates." I nodded and said, "I will take care of it." It had taken me some time, but I was finally starting to get used to working in the type of world that BLM occupied; not one of reality but fantasy!

Of all the programs run by BLM, the Wild Horse Program has been the most visible and controversial. The public imagination runs wild when they see videos of wild horses streaking across the plains, with their heads pointed towards some distant object, and their tails streaming and floating in the wind. People have the sense of what it is like to run "free." The old song, "Born free, free as a bird…" holds their imaginations. I ran into a couple of "seniors" with binoculars several years before I left BLM who had spent the entire summer trying to find wild horses and were excited as kids when they finally found some at the end of the summer. Wild horses seemed to be very hard to find. Not surprising, I guess, as there aren't too many of them!

I got somewhat involved in the program by accident. A roundup of wild horses had occurred in the BLM Craig District in northwest Colorado while I was working in the Meeker office. Not enough employees were available one weekend to babysit the wild horses until some decision could be made on their future. I volunteered for one night because I wanted to see these animals up close. How, I wondered, did these wild horses compare with the ranch horses I had been riding for 15 years during my field seasons with the U.S. Geological Survey? When I got to the location where they were being held, I was very surprised

148

by what I found. All of the animals were in extremely good physical shape. None of the horses had any resemblance to the photo on the wall I had seen in the BLM office.

A wild horse is an animal that is of smaller stature and has a much more muscular and rugged appearance than a ranch horse. Even though these wild horses might be placed up for adoption to the public, they wouldn't be able to be placed with any consistency. Wild horses in other areas, particularly in the states of Oregon and Washington, have had ranch horses finding their way into some of the herds. This interbreeding has produced some beautiful animals. Adoption or placement of those horses would be more productive than the animals I was now looking for.

The attendance of BLM employees at the corral site was deemed necessary by management to prevent some, "Environmentalist sneaking in during the night and opening the corral gates." A dilapidated old trailer had been brought in for the attendant to stay in during his temporary duty. I had thrown my sleeping bag on the old trailer couch for the night as the evening wore on. There were no lights in the trailer, but I brought my lantern along just in case. The night seemed to stretch on forever, with my attention periodically being broken by the field mouse who thought the trailer was his abode. I would hear the tingle of tin as he came through a hole in the back of the furnace. It would take several minutes and then he would stick his head out through another hole in the front of the furnace. Since I didn't want a mouse running around the trailer while I was trying to sleep, I would throw a pillow at him; he would instantly disappear. When I turned off the lantern, he would return in 30 to 40 minutes. I would turn on my flashlight and the pillow would once again be in action. Towards morning the mouse finally conceded defeat.

Nighttime, however, gave me ample time to consider what was going to be the final destination for these wild horses. It was obvious they were going to be sold and end up in a rendering plant, either as dog or cat food. This final act isn't something that BLM will publicize when they announce another wild horse roundup to save the American West. No announcement will be issued by BLM about the numbers of wild horses they disposed of to protect the American way of life. Wild horse sanctuaries are just that — *but only for the very few horses that get that far.* The assertion by some in BLM that, "We only dispose of the very disabled" isn't true as nature itself will make that decision long before a BLM roundup.

When light finally managed to creep into the darkness the next morning, I took several long walks along the corral fence to take one final look at these beautiful animals. It was hard to believe that within several months they would be cut up and put into something that looked like a sardine can. As I continued to look at them, their hides started to glisten in the sunlight as the shadows slowly shifted across the corral. Their features indicated that they were dreaming of running wild once more across the wild plains. Once more, "Born Free," but it was not going to happen.

The problem, as usual with government agencies, lies in the way funding is received from Congress. If money is allotted for the Wild Horse Program, then wild horses will be rounded up so that BLM can say, "See, we are managing the Wild Horse Program." Money will be spent to hire helicopters and pilots to go out and find wild horses somewhere, as well as building fences and other holding pens to use up the money.

Most roundups are taken not because there is overgrazing on federal land, but because of a theoretical "head count." It doesn't matter if there has been a wet cycle and the range is in excellent shape. The roundup will take place because BLM will state, "We believe this action is necessary because the wild horses are causing significance damage to the range." Basically, the BLM is making a political decision to placate ranching interests who claim that the wild horses are destroying the range for their cattle — not one based on reality. Unfortunately, BLM IS BASICALLY SELF REGULATING. If BLM states that a wild horse roundup is necessary, that will take place. There is no "oversight" committee to determine if that program is really necessary at the time. For all practical purposes, BLM has now become God!

The *Arizona Republic* newspaper, in Phoenix, Arizona, reported an announcement by BLM in early February of 2012 that stated, "They had just rounded up 380 excess wild horses in Nevada, with 80 released back in the wild." Apparently, 300 of the horses will be relocated in canned cat or dog food.

The *Denver Post*, in the Sunday edition on September 30, 2012, carried a front-page article and almost two full interior pages on the mismanagement of the Wild Horse program by BLM. As usual, BLM replies to criticism by giving out misleading information. It inflates the number of wild horses to justify the millions of dollars it receives to manage the program. Horses are so few in number, not the 35,000 they cite, that they must use helicopters to go out and find them. The cost of 76 million dollars to place and keep wild horses in corrals and other facilities would be much less *if they didn't round up so many horses to justify the program*! BLM is now selling excess horses for ten dollars ($10) per head. BLM is committing fraud on the American taxpayer by not receiving full market value for horse meat. A 1,000 pound horse would dress out at 550 pounds of meat that would be worth over five thousand dollars ($5,000) on the European market.

The Legislation reported in the newspaper clipping shown above is the first step in allowing horse meat products to enter the U.S. food chain.

AFTERWORD

The only way to resolve the "wild horse count," i.e., the number of wild horses that actually exist, is for an organization to file a Freedom of Information (FOI) request requiring all State Directors of BLM to state the actual numbers of wild horses in each Management Area; including the number of wild horses that are held in other facilities. This count should be verified by an independent citizen's review committee! The current estimate of wild horses by BLM indicate that more wild horses exist than deer or elk in the American West.

Wild horses that will be processed for human consumption are shipped into Mexico, as there are no processing plants for this activity in the United States. Horse meat is shipped to Europe where it can be sold for a high price. Horse meat for human consumption has never been a desirable product in the United States. As soon as plants are approved for processing horse meat in the United States, that meat will slowly work its way into the American food chain.

.　.　.　.　.　.　.　.　.　.　.

On June 6, 2013, the *Denver Post* carried another front-page story on the Wild Horse debacle. This article, like others in the past, has missed the main point about the Wild Horse program run by BLM.

1. The Wild Horse program is used merely as a justification by BLM to obtain big money from Congress for running their different programs. Their justification for rounding up

wild horses is that ranching interests are concerned that wild horses will affect the amount of grass available for their beef cattle. Therefore, BLM will try to make Congress believe that "political pressure" mandates greater funding. Every time a news article is written, the BLM inflates the number of wild horses it is holding to try and justify more money from Congress.

2. The National Academy of Science commentary on the Wild Horse program is accurate in one regard — the BLM doesn't know how many wild horses actually exist on the open range. If the BLM doesn't know how many wild horses exist, how can it make accurate statements about the environmental damage the horses have caused? Of course, the whole program is driven by money and politics. Unfortunately, wild horses get caught up in the process and end up in something looking like a sardine can — they don't have the ability to make political contributions to members of Congress!

Chapter 16

THE UNIVERSE — A MASSIVE SPINNING GALAXY

Time and space — Was Einstein wrong?
You too can journey to the edge of the universe

Science has had some success in trying to explain the origin of the universe. I originally become fascinated by the scientific research on the universe when I was a young field geologist in the early 1960s. Sitting around a campfire at night gave me ample opportunity to gaze upward at the various aspects of creation. The subsequent universal acceptance by science of the Big Bang theory for the origin of the universe has spawned the controversy of whether the universe is a Closed System or an Open System.

Under a Closed System, the universe is not expanding or contracting but is in a stable or "static" condition. Under an Open System, the universe has no boundaries and could literally expand forever. In other words, the universe can expand to a state of being where it no longer exists in 50 to 100 billion years because of the expansion. In the Closed System, there is a controlling element, God if you wish, that would prevent the universe from contracting or expanding — it would always stay the same. No controlling element is present in an Open System.

Einstein originally favored a Closed System. In order to prevent the universe from collapsing because of the gravity, he developed the concept of "cosmic repulsion," an outward force that counteracted gravity. When Edwin Hubble's research showed that the universe seemed to be expanding, Einstein gave up the idea of an outward force, calling it the "greatest blunder of his professional life." My belief is that his concept of a stable or "static" universe was basically correct.

I believe that his "cosmic repulsion" force was really nothing more than the centrifugal force generated by a rapidly spinning universe. Although the literature about a spinning universe is rather barren, there is no reason it cannot be happening. A recent news report indicates that an international team of observers used NASA's newly launched NuStar and European Space Agency's Workhorse XMM-Newton to calculate that high-energy x-rays were spinning in a Black Hole in a nearby galaxy at 670 million miles per hour, or near the speed of light. If it can occur on a small scale, it can happen on a much larger scale. The evidence against an Open System will continue to mount.

Modern science has jumped into bed with the theorists who demand an Open System, with a universe having no limits. They base their evidence solely on the red-shift of various cosmic features. When the wavelength of light increases, it causes a change in color making the pattern more red-shifted, thus indicating a greater speed and distance. The Nobel Prize Committee gave three American astrophysicists $1.4 million for their work in showing the red-shift had increased 5, 6, 7 or more times in the outer portion of the universe. It is extremely unfortunate the Nobel Prize Committee blundered by giving them more credibility for proving universe expansion.

Another way to interpret their findings is that they had actually found the edge of the universe! This is the point where the gravity and outward expansion of the universe reach equilibrium. The universe is neither collapsing nor expanding; another way to state it is that the universe has reached the final expression of its expansion boundary.

A simple way to explain the bending of the universe is to take a long piece of fairly rigid paper and make a series of even folds the entire length. Place this folded rigid paper on a flat table with one end of the paper folds extending off the table. Now, start bending the portion that is not on the table and also start pulling gently to the end. You will discover the folds start expanding. This enhanced bending mimics the lengthening of the light wave. The bend that the three astrophysicists found would indicate a very severe bend, one that would occur around a very flat universe.

Both the Closed and Open Systems demand the Big Bang as their origin. The universe was born in a titanic fireball that was the creation of everything; all space, matter, energy, and even time itself came into being. It is hard to realize that there was no sky, clouds, light, air, moon, sun, starts, people, or animals in the beginning. The Big Bang was affirmed by the discovery in 1965 by a background of uniform microwave radiation that extended throughout the universe. The microwave background was produced by matter cooling off after the fireball. Even the Catholic Church in 1982 agreed with the concept of the Big Bang.

The concept of the shape of the material in the incipient fireball is very important in considering how the universe developed. The original fireball was supposed to be a point of origin, a round ball of something because the strength of the cosmic microwave radiation seemed to be uniform in all directions. Conjecture, however, does not equal knowledge.

Because science does not recognize a deity, this conceptual model of the universe beginning has never seemed important in any discussion of the universe. But, suppose the original matter in the fireball was not quite round. Let us assume that it was a spinning, slightly oblate spheroid with a slight bulge on its top and bottom. Suddenly, the "ripples in the cosmic microwave background radiation" that produced the "hot spots" and "cold spots" that astronomers are so concerned about are no longer a mystery.

If various shapes for the beginning of the incipient fireball are drawn, ranging from a spinning oblate spheroid to a spinning disc, projections of the original energy will show how various clusters of galaxies could have formed. The spinning oblate spheroid configuration will produce background radiation similar to that produced by a point of origin. It is interesting to draw many configurations and see how the energy would evolve in the universe.

Any current description of the origin of the universe does not take into account the spinning nature of the universe which evidence indicates certainly exists. Recent evidence from Europe indicates that the "expansion" is slowing down and that the universe is slightly older than before, now 13.8 billion years old. However, when the "spinning" nature of the universe is finally acknowledged, it will be discovered that the universe is much younger and smaller than previously believed. The expanded light wave is distorted by the spinning, thus giving the illusion of a much bigger expansion.

Although science would like to believe that it is always searching for the ultimate truth, it sometimes gets stuck with certain dogmas like political parties have experienced in the past history of this country. If a party believes it is liberal or conservative, then it automatically believes it has to expound on certain values that are deemed necessary to uphold those values. There is no thinking "outside the box" to ascertain whether those values are still actually true. Once the concept of a round point of origin for the beginning of the universe is considered to be a fact, it tends to perpetuate itself.

Years ago, I played in a simultaneous chess exhibition against Petrosian, Russian Grand-Master and former World Chess Champion who was touring the United States. There were 25 chess players grouped around tables which faced the center of the room. Petrosian would stop at one chess board, look at it for several seconds, make a move, and then proceed to the next chess board. This procedure continued all night. The Grand-Master could remember every chess position on every board throughout the evening as the games wore on.

As players dropped out, I was one of four or five players who survived until the bitter end. What finally happened? I went for "theory" eventually without really looking at the position my chess pieces finally occupied. We should considered that the "red-shift" of far away galaxies does not automatically mean that they are flying away into oblivion!

Theory does not equal fact!

. . . .

There will always be different theories that will try and explain the universe. One of the latest versions considers our concept of what is "existence." This theory involves the concept of a hologram and how it relates to the universe. We see an example of a hologram every time we look at a Master Card or Visa credit card. It is the small shiny image that gives a 3-D visual affect from a flat surface. We obtain a 3-D image from something that shouldn't be there — we are observing a projected image. All the data is encoded on flat surface.

This theory would take a flat surface, like a flat universe, and project everything in existence like a hologram. All data, all information, everything that is to be and will be in existence can be observed. Doesn't the Bible say that God will make the whole world his foot stool? Under this theory, we are not too far from reality; especially if we combine this theory with a flat universe that is much smaller and younger than current science believes.

. . . .

My interest in the universe really started one night in 1962. D.J. and I had just put in a "horse camp" high in the upper reaches of the Snake River Range in southeastern Idaho. The first night our evening fire had slowly burned down as darkness settled in around us. We had made a fire in a partly wooded area next to our tent. A yearling doe had decided we were a curiosity she could no longer ignore and continued to edge closer to the fire. She would disappear for a few seconds, and then mysteriously reappear. Finally, she edged up to an axe stuck in a log next to the fire and started licking the handle for its salt content. A slight movement by us sent her scurrying away never to return. We edged closer to the fire as the cold continued to settle in around us. Even the insect noises had now faded away.

The night sky now blazed above us, as the Milky Way suddenly started to put on an unusually spectacular show. There were no light sources from the surrounding area to lessen the effect of being surrounded by complete, utter darkness. The last faint glow of our fire finally flicked out. The night sky now seemed to come alive with much anticipation — and then it began.

At first, there was only an occasional shooting star, winding its lonely way through the cosmos. Suddenly, there were many more, lighting up the heavens, the last vestiges from an unknown origin that was now making itself known. The night sky continued to dance to the delight of the heavens as space debris decided it was time to come home to its final resting place — planet earth. As I sat on a big log watching this amazing display, I started to think of part of an old nursery rhyme I knew as a young child — "the little dog laughed to see such sport, and the dish ran off with the spoon."

As I tossed and turned in my sleeping bag that night in 1962, I realized that no rational explanation yet proposed could account for the amount of space debris currently found in this solar system. Despite years of scientific study, something was wrong with our approach to understanding our universe. Closer to home, science couldn't even come up with an acceptable, rational explanation for the origin and existence of our moon! What was an explanation for what we see and experience in our existence?

Chapter 17

THE EARLY YEARS

They say I was born in an old wooden bedstead that stood in the lower bedroom of our farmhouse in south-central Wisconsin. I don't rightly remember the occasion, but everybody says it was true. You can't argue much with witnesses, especially when there are several of them and they all agree. The doctor, nurse, and my mother all declared later it was a much awaited "coming out" event, one that I would treasure for the rest of my life. Perhaps as I get older, I will remember more of this momentous occasion.

The above paragraph illustrates how Mark Twain would probably have written a description of his birth if he had been born in 1936. To us, his writing now seems "quaint." Yet the world I initially grew up in resembles that of Mark Twain much more than the modern society of today. My sister was born only 16 years after the death of Mark Twain in 1910. Since the time of my birth, the following changes have occurred.

- Ballpoint pens
- Microwave ovens
- Cordless tools
- Cellular phones/ DVD's/CD's/iPods
- Flat-screen digital TV sets
- Statin drugs
- Open-heart surgery
- Polio vaccine
- Pull tabs on aluminum cans
- LED lights
- Hearing aids
- Lasik eye surgery
- Artificial joint replacement
- Ceramic kitchen knives
- Solar panels
- Electric blankets
- Pantyhose
- Wireless doorbells
- Stent implants
- Penicillin
- 8-track tapes
- Tube TV sets
- Computer chips/ computers
- Interactive computer games
- Electronic chess sets
- Polaroid cameras
- Digital cameras
- Computer printing of newspapers
- Microsurgery
- Laser beams
- Walking robots
- Airbags on cars
- Disposable diapers
- Internet
- Interstate highways
- The Hubble Telescope
- Space stations
- Jet planes
- Rocket propelled engines
- Global positioning systems (GPS)
- Synthetic oil
- Hybrid cars
- Space satellites
- Moon exploration
- Space probes
- Helicopters
- Satellite radio

- Pilotless war drones
- RADAR
- Smart bombs
- Nuclear submarines and carriers
- Artificial heart valves
- Atomic energy
- Atomic bomb
- Jumbo headed golf clubs
- Social Security
- Etc., etc., etc.

It is very difficult to describe my early years because of the great differences that existed at that time from today due to the economic depression and World War II. During the 1930s men would come out to the family farm and beg for work from my father, asking only for a dollar a day plus dinner. My mother also traded chicken eggs to the local grocer as payment (egg money) toward her grocery bill. Bags of flour were bought in 100-pound cotton bags that had brightly colored designs; these bags were then made into clothes for my mother and sister. A woman teacher for the local country school stayed with my mother for part of World War II and paid rent with her sugar coupons. Letters could be mailed with 3-cent stamps. Everybody was poor, but nobody knew it!

The family was fortunate to be living on a farm during this period because we could grow and can most of our own food. Every year, a hog was butchered on the farm and put in our big freezer. The fat was cooked on our wood stove in the kitchen and made into lard. The lard in those days was used in the making of everything, cookies, bread, cake, pancakes, etc. If the lard wasn't used for baking, it was used to fry everything at meal time. Potatoes were grown and stored in a big bin in our basement. My mother also had a big garden. We had as much milk as we wanted, and drank it even though it was not pasteurized. I could never stand the taste of milk as a kid and only could be persuaded to drink it by putting huge amounts of Ovaltine in it; one count showed I had consumed well over 120 cans of Ovaltine by the time I was a teenager.

The dairy farm was pretty typical of farms during that era. There was an old rambling two-story house, dairy barn, chicken house, hog house, corn cribs, tobacco shed, granary, and a somewhat garage. Electrical lights had been installed in the house and barn only a few years before I was born. A windmill next to the house supplied water not only for the house but also for the dairy barn. The house had an old dark damp leaky cellar underneath it. This cellar contained a big bin for storage of potatoes and another large bin for coal. Coal was slid into the coal bin by using a small chute through a cellar window.

The laundry was also located in the cellar and consisted of a single one-speed machine that had a wringer on top of it. After the clothes were washed, they were put through the wringer and then taken outside and placed on the clothes line. My mother usually made white lye soap for use in the family's laundry. In another corner of the cellar was a fairly big open cement enclosure called the cistern that held water which was then drawn up to the kitchen by a hand pump. Lining the walls of the front part of the cellar were rows of shelves, all holding a large number of glass canning jars containing the results of the summer garden. The cellar contained no furnace for heating the house during the cold months, as this was done by wood/coal stoves on the first floor.

Family farm in 1958. This house is typical of the farm homes found in the area.

The house had a main entrance that led directly into the kitchen with a big wood stove that held a large oven immediately to the right as you entered, and a table and the refrigerator to the left. The main kitchen sink with a hand pump was directly across the room. In the middle of the 1940s a lean-to was added to the house, with the doorway just to the left of the sink, which contained a square bath tub, sink and toilet. This was the first indoor plumbing that existed in the house. An outhouse or "privy" had been used until this time, even in the middle of winter. Water was heated in big gallon containers on the wood stove and poured into a laundry tub on the second floor for the weekly bath sessions on Saturday night.

To the right of the refrigerator was a doorway that led into the dining room, which held an old oak table. To the left as you entered the dining room were two doorways. The first led to the bedroom in which my siblings and I were born. The second doorway had double doors which led to the living room or "parlor." The doors into this room were usually kept closed during the winter months and seemed to be only opened for company during warmer months. The room held the piano and several upholstered chairs and a formal couch. It also had a rug in the center of the room. Between the two doorways was a big upright coal stove that heated the house during the cold weather.

Directly across the dining room from the kitchen door was a steep stairwell that led to the second floor. A small bypass of the stairwell led to the steep open steps into the cellar. At the top of the stairs you either turned right or left. To the right was a fairly big bedroom that was heated by the stove pipe coming up from the coal stove below in the dining room. My father and brother usually slept in this room. To the left of the stairwell was another much smaller room called the "honey" room, and beyond this room a much larger bedroom. My sister usually slept in the smaller room, with the bigger bedroom closed off during colder months because there was no way to heat it. For the first few years I slept in the downstairs bedroom with my mother during the winter months. When my sister graduated from high school and moved to Madison, Wisconsin, I inherited her room. The bed was heated by "flat irons" in winter, which were heated on the kitchen stove and placed in the bed wrapped in towels under the quilts.

Most meals were eaten in the kitchen with the dining room usually reserved for guests and holidays. If we didn't have visitors on Sunday, the family usually had a traditional "fried chicken" dinner in the dining room. As I got older, it became my job to catch the chicken! The table could be vastly extended in the fall when we had "thrashers" in during the fall harvest. My earliest memories of the dining room were zooming around with my "walker." It was made with two round hoops, with four wheels on the much larger, lower hoop. The upper and lower hoops were connected by four pieces of strap iron bent to accommodate the different hoop sizes. The upper hoop had four leather straps that held a small wooden seat. I probably was around 10–14 months during that time of my childhood.

Even though the late 1930s and early 1940s could now be considered to be of an ancient age, our family did have two technology-wise useful conveniences. One was the old wall phone which hung on the wall next to the stairwell. It was the "party line" type, meaning that there were four to eight customers on it. Each party had its own set of rings; for example, for one it might be two short rings and then one long ring, or one short and two long rings for someone else. It was always a guessing game trying to figure out who else was listening to your conversation. Usually you ended your conversation and held the receiver tightly to your ear trying to count the number of "clicks." To make a call outside the party line, you gave the handle a long crank to summon the telephone operator.

The second convenience was a small table radio in a corner of the dining room. As a young boy, I was addicted to listening to the adventures of "Jack Armstrong, the All American Boy." These adventures were usually in sequences that might last several weeks. Therefore, when the program came on late in the afternoon, I usually had the radio turned down with my ear next to the speaker. I knew that as soon as my father came in for supper, he would want to hear the latest news about World War II, and I tried to not let him know that the radio was on. How could anyone equate Jack Armstrong trying to escape from the "Lost World of the Dinosaurs" with Gabriel Heater or some other commentator?

As you stepped outside the kitchen door onto a small flat concrete porch, you got to survey the rest of the farm buildings. About 40 feet away and left was the windmill. Immediately beyond that was the chicken house. In back of the chicken house was a small building called a brooder house. This small building was used to hold about 100 week-old chicks that my family purchased early in the Spring months from a hatchery near Stoughton, Wisconsin. These chicks were placed under a small round heater for warmth, but some losses still occurred. Later on, when chickens were no longer raised, this small building was used one summer to raise a small pet skunk we called "Honeysuckle." We never had this skunk "neutralized" and eventually it returned to the wild. My brother got a big kick out of taking this skunk when it was small into a tavern and putting it on the bar. This skunk had a nasty habit of running backward at you with its tail raised!

Directly in front of you as you stood on the porch was the dairy barn about 50 yards away, with a silo fronting towards the house. The upper level of the barn was mainly for the storage of hay. Hay was taken off the hay field loose, and was unloaded by using a big rope sling. I got the job of spreading the loose hay when I got older. It was usually so hot in the hay loft that I would take two bath towels with me to deal with the humidity. If it was 90 degrees outside the barn, it was probably closer to 120 degrees or more in the hay loft — and dusty!

The barn and I ended up having a very special relationship. When I was 16 years old, it was decided by the family that the barn needed a new coat of red paint. For some reason, almost all barns in the State of Wisconsin were painted red. The prevailing opinion was this originally happened because red paint was cheaper than any other color. Occasionally somebody would paint a barn white or even blue, but this was very unusual. The family decided it was too expensive to hire a painter to paint it by hand, and spray painting was quicker but did not cover very well. But Marvin, however, just happened to be available.

So I got the job of painting the barn, a job that lasted close to six weeks, and eventually involved putting 25 gallons of

This photo was taken in 1952 when I was 16 years old. Note that only the part of the barn to the left of the silo has been painted at this time.

red oil-based paint on the barn by hand with a four-inch paint brush. With wind and trying to work on a 30 foot wood ladder for higher sections, it became a question of what got the most paint. Oil-based paint will splatter quite easily and frequently I was literally covered in red paint. After every session of painting, I was forced to use paint thinner to clean off my glasses, face and sometime my hair. The hardest part of painting the barn came when I had to paint the front peak. This involved nailing wood boards on the lean-to roof and placing the foot of the ladder against these boards and leaning the top of the ladder against the peak of the barn. It was very scary with the hot summer sun, flies, some wind, sticky paint spatter, with the painting being done 40 feet or more from the ground surface using a rickety old ladder. Luckily, I survived.

After a visit to the old farm homestead in 2012, it was evident that the barn had suffered neglect for a long time. The roof had caved in, but a few rafters were still there. Most of the side boards had fallen inward. The cow stanchions had been removed. Weeds and small growth had enveloped the sides of the old barn covering most of the old cement block walls. The silo still stood, a stark reminder of olden days. So like all things, the barn had finally come to the end of its useful life; but the memories still remained. My brother and I had mounted a basketball net at the far end of the hay barn and many a night and Sunday afternoons during the cold winter months were spent shooting baskets and playing "horse." This was a game where one would make a shot from a certain spot and dare the other to make it from there too! Of course, the memory would also remain of spending some very rainy summer afternoons sacked out with my dogs in the hayloft. The unpleasant memories of long hours spreading loose hay and hand milking cows in the very hot fly-infested humid summer months had faded away a long time ago.

. . . .

The round Egyptian-like monoliths called silos found around most dairy farms are built to hold silage, which consists of green corn and stalks chopped up and blown into the silo for winter feed for the dairy herd. Cows are normally dry and have little or no milk production in the summer months before they produce their yearly calf in the fall and resume their milk production. Consequently, no extra feed is usually given to them when they can obtain almost all their feed from being out in the summer pasture.

Silage is normally fed to the dairy herd in the morning along with a balanced grain supplement when they are not out-doors. Usually enough silage is obtained through the fall harvest to carry the dairy herd through the winter months.

When not enough silage was available one spring, we had to buy additional silage from a local farm near the town of Deerfield. This produced one of the most bizarre happenings of my young life. Most silos are not attached to the barn proper, but are connected by a small enclosed passageway. This allows silage to be thrown down from the silo into this small enclosure where it can be picked up and fed to the cows. The first time we went to pick up some of this silage, we unhooked an old wooden door on this enclosure, and suddenly had rats, rats and more rats running around this small old room. There were big rats, small rats, brown rats and more big rats! Some of them seemed as big as a cat! Apparently, the corn in the silage had fermented and the juice had seeped out and produced a molasses like substance on the floor which the rats were feeding on. This certainly would account for their big size.

The first time this happened, it was a complete surprise to both my brother and me, as we were unprepared. Next time we came prepared with a couple of axe handles. We would sneak up to the door and quietly and quickly open it and start swinging our axe handles at the rats. The rats would go absolutely wild, squeaking as they ran around trying to climb back through their original hole to the barn, while others were trying to climb up the walls, and also trying to get through the door past us. We usually would manage to get several of them each time we were there.

Although this scene was repeated almost every time

we went to get silage, one time something was different. We had gotten so used to the rats running away, we never gave any thought that we could be a target of their escape. Opening the door, we had the usual explosion of the rat population, but then the unthinkable happened — one ran up my leg underneath my blue jeans. I could feel the tiny claws digging into my flesh as it clung to my lower leg. I put both hands around my knee so the rat couldn't continue to climb. My brother still stood there, swinging his axe handle at the rats. "Keep swinging," he yelled. "I can't," I replied. "Why can't you?" he asked. I said, "Because I have a rat hanging on my leg." He just stood there thinking I was a complete idiot. I still didn't move, trying to be as quiet as I could. It seemed like an eternity. He said "I don't see any rat." Finally, the rat decided he couldn't go anywhere and I felt the tiny claws run back down my leg as it ran away. My brother just stood there and finally said, "I see what you mean." "Thanks!" I said.

. . . .

The main entrance of the barn was on the south side of the lower level. Just outside the entrance was a cement water tank for the dairy herd. As you entered the barn, a storage room was on the left and an enclosed room on the right which held a cold water tank for holding the 10-gallon milk cans from the morning and evening milkings. As you entered the main part of the barn, a row of stanchions for holding the cows while being milked extended along the right side of the barn. The immediate left as you entered the barn had a large pen for young livestock. A walkway divided this pen from the next row of stanchions on the left side of the barn. The walkway allowed access to the silo. The horse stalls and bull pen were on the far end of the barn.

Milking was accomplished by using a milking machine connected to a vacuum line run along the top of the stanchions. In the spring when cows were becoming "dry," i.e., not giving much milk, milking was often done by hand. When the rainy season hit and the cows were very muddy, and the flies were bad, it was a constant battle to do the milking. Cats would always lineup to have milk squirted at them. They would open their mouth to hope a stream of milk would be coming. Often, the cats were covered with milk. I made the mistake as a teenager when I was faced with this constant battle with the flies, to tie one of the cows' tails to the one-inch vacuum line to limit the amount of tail switching. Unfortunately, I forgot to untie it when I turned the cows loose after milking. I can still see and hear the 40 feet of vacuum line being snapped off its supports by the cow as it followed the other cows, dragging the entire line to the outside cow yard. As my brother was rather handy with tools, it only took three to four hours to repair!

The pig house and its outside pen were located next to the outside water tank. Two corn cribs were next and were connected by a central roof, with the inner space allowing a corn wagon to be driven between them. The cribs were filled by hand with a scoop shovel — a rather tedious process. The corn crib next to the pig house gave excellent cover for a Sunday afternoon session of shooting sparrows off the pig house roof with a single-shot .22 rifle. Sparrows were a constant problem, flying down to steal pig feed out of the feeding bins and wasting a lot of the feed. The granary was directly across from the corn cribs, followed by the tobacco shed and garage coming back towards the house along the small entrance road. The small outhouse sat next to the garage. A small gravel road led down to the barn entrance from the main county gravel road on which the milk truck came down every morning to pick up the milk cans. There was no mistaking when the milk truck arrived as our dogs would start chasing the truck, barking furiously and trying to bite the rubber tires.

. . . .

There is nothing glamorous about growing up on a farm, or the farm work itself. Although Mark Twain had his Mississippi River adventures to occupy his memories, it unfortunately did not flow through south-central Wisconsin. I grew up the youngest of three children, my sister Caryl being ten years older, and my brother Edward five years older. Being the youngest did have some advantages. I missed most of the depression years, and I was born too late to serve in either World War II or the Korean War. After college, I was declared 4-F, meaning I was exempt from military service because of a trick left shoulder, i.e., one that could come out of its socket when the arm was pulled a certain way. The shoulder had become dislocated 30 to 40 times between my senior year in high school and my senior year in college and constantly gave me trouble until the mid-1960s.

After I entered the University of Wisconsin as a student in 1954, I was required to be in the Army ROTC program for the first two years. I can't say I was thrilled at the opportunity. I was required to attend a class every week for the

My brother Edward and sister Caryl in 1948. He is five years younger in age but now taller than his sister.

ROTC program. At the end of my first college year, my ROTC unit was paraded up Main Street to the Capital in Madison, Wisconsin, to take part in Memorial Day Services. Unfortunately, we had to march in formation in wool uniforms carrying our rifles and field packs. The heat was oppressive, the temperature well over 90 degrees. Every so often a cadet would keel over from the heat. At the end of my second college year, all ROTC units were marched into Camp Randall Stadium. All units, wearing wool uniforms and carrying rifles, had to stand AT ATTENTION in formation facing into the late hot afternoon sun. The officers reviewing the ceremony all sat in the shaded part of the stadium. Students were dropping over like flies in the hot stadium.

Consequently, when the push came to sign up for the last two years of ROTC, I decided I had enough of the program. If by some chance I had agreed to take the last two years of ROTC, I would have graduated as a Second Lieutenant in the U.S. Army. If by some chance I flunked out of the program, I was automatically drafted into the Army. If again I continued with the ROTC program, I would have to serve three years. No matter how I looked at it, I didn't consider it to be in my best interest, as I had already decided I wanted to go to graduate school for a master's degree.

Right after I graduated from the University of Wisconsin, I received my notice to take my physical for possible induction into the Armed Services. At the appointed time I was one of twenty young men lined up naked to be inspected. At that physical, I informed Army personnel of my disability. Their retort was simply, "If you're lying to us, you will be inducted tomorrow." Within a week I was transported by myself in a 40-foot Army bus to Milwaukee for additional medical tests. After my return home, I received a letter ten days later notifying of my 4-F classification.

. . . .

My father had served in the First World War as a private in the infantry. At the age of 26, he was drafted and after a short boot camp, he was loaded on a troop ship with 5,000 other draftees and sent overseas, landing in Brest, France. He was projected to be moved into the front lines within several weeks. On the night before his unit was to head for the front lines, he came down with the mumps. Consequently, he was quarantined for three weeks. When he had recovered and was declared fit for combat, his unit no longer existed! A position was then found for him guarding and transporting German prisoners of war, his being well

My father and mother standing in front of my grandfather's house in Deerfield, Wisconsin during the early 1940s.

qualified having grown up speaking German every day. His only fond memory of his entire time in the war was simply that the German cooks were some of the best in the world!

When my father returned after the war, he briefly

worked on his father's farm on Highway 73, east of Deerfield, Wisconsin. He was determined to eventually have his own farm even though he only had a 6th grade education. He had been taken out of school when he was 12 years old to cut wood and help support his family. It was at this time upon his return from the war that he met my mother, who was teaching at the local Old Deerfield grade school several miles from the Schroeder farm. While my father was of German heritage, my mother was born on a farm near Cambridge, Wisconsin, a Norwegian community. She had taken three months of college after high school, enough education at that time to obtain a teaching certificate. Their marriage in 1921 led them eventually to buy the old Bannon farmstead, where my siblings and I were born and grew up.

The war for my father did have some bad effects. When he returned home from the war, he found that the Schroeder farm had been promised to one of his older brothers. He had spent most of his childhood working on that farm, and he felt that he should have had the chance to buy it, especially when he was in the U.S. Army fighting for his country. Consequently, an estrangement developed between him and his two older brothers. When I was growing up, I never met his older brothers, nor did I ever meet his brother's children. His main relationship and best friend was with his younger brother Henry, who lived in the Stoughten area.

. . . .

Education was something that my parents believed in deeply, with my siblings and me all finishing high school.

I was the only one who eventually went on to college, and then graduate school. We all usually walked to the local one-room country grade school that was about one mile from our house. I started to use my bicycle when I got older. The school housed only the first through the eighth grades. Only about 18–20 students attended this country school at any one time. I started out with three to four classmates but, by the sixth through eighth grade, I was the only student in those grades.

The school house was made of brick and was very well built for its time, probably better than 95% of the farm houses in the community. One climbed a double-sided porch to enter the school. There was a small mud or coatroom to leave coats before entering the big school room. All the small desks faced the front of the room, where a long low table was placed. The teacher sat at his small table to conduct classes. Since all classes were taught in front of all students, by the seventh and eighth grades, you got pretty adept at ignoring the teacher and her lower grade classes. In the back of the room was a big upright wood stove that heated the whole room during the cold weather. My mother would send along a lunch that could be placed on this wood stove to give me a nice "hot" lunch — quite a treat over cold sandwiches. The wood for the stove was stored in a small building near the front entrance.

Recess, of course, was the most fun time. In the winter we poured water on a small hill next to the school building. It would freeze over and provide a very slippery surface which we would run at and then slide down on our feet. In the warmer months the younger children played on the swing set while the older kids played anti-over the school building with a tennis ball. The object here being to catch the ball and immediately run around the building trying to hit another kid. Great fun! Another game was called pump-pump pull-away. This consisted of two lines of kids, facing each other across an open space, daring somebody on the other side to catch us we tried to cross to their side. If touched, we became their prisoner. We could only get free if somebody from our side could rescue the prisoner.

With the school situated on the north side of a fairly big wooded area, field trips would also be made throughout the wooded area collecting nature's treasures. If kept inside during bad weather, there were always a number of silly card games to be played. The first few years of grade school we also had Christmas programs in which all the kids participated. These were discontinued by the time I reached the upper grades when we had a different teacher. The worst time during the school day was using one of the outhouses, especially during the winter months, as there was no indoor plumbing.

One could say that I was at an immense disadvantage by having to go to this small country school. My world, however, extended way beyond the confines and four walls of this school. I attribute this to the small but quite adequate library. Probably five feet high by eight feet long, it contained a great variety of books, and I read most of them. When reading my world would suddenly start consisting of going on a safari in East Africa with Osa and Martin Johnson as they photographed wild animals; or Frank Buck as he trapped wild animals in Borneo to bring back to the zoos of North America; or with Jim Bridger as he fought with the Indians while trapping beavers for their fur in the wild west; or panning for gold by prospectors in the mountains of South America; or being a White Hunter on the trail of a man-eating tiger in India. These were all the day dreams of a kid taking him out of reality. Little did I know at that time that the first 20 years of my professional career would rival any of those wild adventures. Somewhere — someplace — another kid with access to a library is starting to have his day dreams too!

While school was a big influence on my young life, the church also played an important role. My family always attended a local country Lutheran church associated with the Missouri Synod. The doctrine or philosophy followed by this church for the time we attended was rather severe. The congregation was divided during the services — the women sat on the right side of the church and the men sat on the left side. The services were long, lasting up to 1 ½ hours. The pews were hard oak, and I never ever saw a cushion used in any of the pews. Dancing and women wearing lipstick were two activities frowned upon by the church. My father served as treasurer of the church for six years. With only six years of education, he was very poor at keeping books, and this duty was taken over by my mother.

When I hit the age of ten, church activities got a little more serious as I started going to parochial school during the summer. I went to school with seven or eight other children in the church basement for three months between 9 a.m. and 12 noon every day. Confirmation in front of the church took place after this three year period and was always a big event. I managed to memorize a great deal of this teaching and could stand up for long periods and quote chapters of the Lutheran catechism when I was asked questions during exam time. My fondest memories of the church events as a child were the Christmas Eve Services. A big evergreen was cut down and children got to help decorate it. After Christmas Eve services, brown paper bags were passed out to the children filled with peanuts, nuts, candy and other goodies. Other memories were of church potlucks in the summer months with tables groaning with food.

. . . .

Entertainment during the war years was partly provided by Saturday night movies in the Deerfield City Park during the summer months. A big screen was set up and a movie projector showed Western films in black and white, usually starring Roy Rogers, Gene Autry, Tom Mix or several other cowboy stars. The sound was usually pretty bad, but the action on the screen was usually pretty clear. There was a round, wooden, bi-level bandstand in the middle of this small city park and it was usually open and sold popcorn for a nickel. Band concerts would usually be performed for the 4th of July celebrations and hamburgers, pop and potato chips were also sold.

The township also held a Deerfield Festival during the summer months. The big event was usually a baseball game during Sunday afternoon in the city park, with many carnival rides present the whole weekend. The local high school gym had a number of exhibits of food, crafts, clothing, as well as farm exhibits of field corn, oats wheat, etc. One time I had nothing to exhibit, so I wandered over and got some of my neighbor's wheat to show and received second place?! During the evening hours the beer tent was usually very well attended.

. . . .

My father's parents lived in a house in the town of Deerfield about four miles from the family farm. My parents frequently visited there during Sunday afternoons. Card games were usually played, but I was too young to play and only watched. Frequently, I would get 15 cents and walk several blocks down to the local drug store and get a big chocolate sundae. Ice cream cones were also a treat and cost a nickel. My grandfather had immigrated to the United States from Germany in 1880 when he was 21 years old. He was the son of a Prussian army officer and could remember polishing his big black boots. Apparently, both of his parents died when he was a teenager. There were rumors that he wanted to avoid the aftermath of the Franco-Prussian war in Europe by moving to the United States. I had walked home from grade school one afternoon when I learned he had died that morning from a heart attack at the age of 85 while peeling potatoes out of his garden behind the house. He had left a long heart-felt letter to his children to be read after his death, as he apparently had experienced some heart problems in the past.

. . . .

The environment that I grew up in didn't allow for much time to play. I started doing odd jobs at a young age such as going into the chicken house searching through nests after fighting with the hens to gather the eggs. I also learned how to put the ears of corn through the corn-sheller in order to feed the chickens. In the winter months I did have some

186

fun using my sled to slide down the driveway after a heavy snow. Of course, every boy has his dog and mine was Sandy, a shepherd that had inherited the collie look. He was my constant companion in those early years. There were no playmates from other farms as they were too far away and

In 1944 I was eight years old — and Sandy was my dog!

they would have their own chores to do. During the summer months, heavy rains would set in for several days and limit any outdoor activities. Some of those would allow me to go into the haybarn and burrow into the loose hay for part of the afternoon — and pull Sandy in on top of me to keep me warm. He was always glad to do this!

As I got older, I got involved in a lot more adult activities. I went along with my father when he hauled hogs and cattle in the back of our old farm truck into Madison to be sold at the local market. I also helped with the planting and harvesting of tobacco, but this activity only lasted a short time. Potatoes were a major crop for the family, as it was a very stable part of our diet. I started picking potato bugs off plants, getting one penny for every five bugs, and finally progressing later when I was older to helping dig up the potatoes and storing them in our cellar. Fishing was done on some weekends with cane poles at some of the local streams, with only carp and bullheads being caught. I went trapping one winter as a teenager and caught 10 to 12 muskrats and one mink. I received $20.00 for the mink fur and a dollar for each of the muskrat pelts. This activity only lasted one fall as the price went down, and I didn't like getting up to run the trap line. Of course there were always the farm animals to be fed, the cow barn to be cleaned of manure, corn to be shoveled into the corn cribs at harvest time, corn stalks to be picked up and chopped to fill the silo, etc.

For some reason, I also started raising white rabbits as a teenager and eventually ended up with over 120 rabbits as they seem to reproduce rather quickly. Unfortunately, there didn't seem to be much of a local market for them.

My mother insisted that I butcher several of them and she took them into one of the grocery stores in Madison and sold them for $6.00. However, I did not want to butcher these beautiful animals and eventually had a trucker haul them to a market in Milwaukee when I was going to go to college. For all of these rabbits, after I paid the trucking fee, I received $12.00 total.

Since I had no way of getting an income or making money as a teenager, I decided to try and raise a bull calf once I entered high school. This calf was obtained through the Future Farmers of America (FFA) program. This too failed to gain me much money, as in the drawing I received a young bull calf that was a twin, plus had a poor pedigree. After taking care of this calf the whole summer, plus a great effort of showing him at the County Fair, I received a total of $120.00. Other kids had gotten much better animals and had received between $400.00 and $600.00 for them. I calculated that I probably received about 10 cents per hour for all my efforts.

When I became high school age, I went hunting in the local woods during the summer months with my dog "King," as Sandy had now died. King was a much smaller dog than Sandy, probably only weighing around 35 pounds. He would run through the woods making a lot of noise and then stop to listen for squirrels. When he spotted one, he immediately chased it up a tall tree, usually a big oak. I would sit down and try and pick off the squirrel with my single-shot .22 rifle. With only iron sights, and a moving squirrel, I usually wasted a lot of ammunition. There is nothing better than stewed wild squirrel meat simmered for several hours on the stove and then served over mashed potatoes.

My hunting instincts were always with me. When my brother and brother-in-law went pheasant hunting one summer, they gave me a single-shot 12-gauge shotgun to use. Of course they persuaded me to go down through the corn field with them stationed on the other end. They expected quite obviously that I would chase up a pheasant and they would get to shoot it. When I got close to the end of the field, I heard a lot of yelling and then shooting, and then suddenly the pheasant rooster came flying by me. I fired my one and only shot and down went the pheasant. It was later determined that I had hit it in the head with a single shot pellet.

About this same time, I made several English long-bows as well as the arrows. I had absolutely no luck hitting anything and decided I needed something better. Consequently, I bought a commercial cross-bow along with the arrows. I set up several targets and spent hours practicing hitting them. I quickly learned to never shoot at a squirrel in a tree as the arrow would take off and if I managed to find it, it was usually several hundred yards away. Finally, I turned to using straight wooden dowels, and they would go straight for 30 to 40 feet and then go off line; but they were easy to find. Squirrels in the local farm trees were always a target, but I don't remember hitting one. The squirrels quickly decided that there were better places to spend the afternoon.

My father had his favorite pump-action 12-gauge shotgun, and if anything was to be shot, he would haul it out. The family joke concerned the time he decided one afternoon to try and shoot a Canadian goose. Flocks of geese were forever flying over the farm in the fall of the year and would frequently set down near one of our corn fields. When he tried to approach a flock, the entire flock would start their honking and fly away. He noticed that whenever he was out with his manure spreader and horses, he could approach them within shooting distance. So one afternoon while spreading cow manure, he took along with his 12-gauge shotgun. Consequently, he did get close enough to shoot one, but the horses bolted from the noise and threw him backwards from his seat into the middle of the cow manure. He always insisted it was worth the goose — but there was some discussion about that.

. . . .

Finally, the time had come. I had to give up my country school ways and enter the big time — High School! The

Deerfield High School in the early 1950s.

change from the country school to the urban high school went off without too much of a change in the routine. Now however, I rode a station wagon with five or six other high-school-age kids. The big yellow school bus was not used on my route because it was not convenient to use in my area. It was not difficult to get adjusted to high school as my class consisted of only 18 students — about the same number of students in my entire grade school. The entire high school had only about 80 students in all classes.

My study time in high school was about the same as what is taught today — English, Math, Social Studies, History, etc. My main outlet to get away from the boredom was to take four years of Ag (agriculture) classes. This involved the feeding and care of farm animals, as well as the raising of crops — everything I had been involved in all of my life. Some of the activities in Ag involved going to Madison as part of a judging team competing with other high schools in the judging of dairy cows and farm crops. Our high school usually finished in the top 10 out of 120 entries.

My other activities consisted of playing my clarinet in the high school band for various school activities as well as being in the marching band. My main sports activity was basketball. For three years I had suffered being on the "B" basketball team because I was not talented enough to play on the main high school team. Between my junior and senior year, I put up a telephone pole in the summer time with just a hoop and a net and spent all my spare time practicing. The pole was set partly on a slight slope, so if you missed the shot, you had to chase the ball down the hillside. It is surprising how good you can become when you are sufficiently motivated! I did make the main high school

The Deerfield High School Basketball team of 1953-54. I'm in the back-left of the photo.

team in my senior year as I had developed a pretty good jump shot — the most points I ever made in a game was 21 points. It was while playing basketball my senior year that I first developed my "trick" left shoulder that plagued me for a number of years. Other sports I tried included running the low and high hurdles as a senior and some boxing in physical education with a kid who was entering the golden gloves competition. I never tried football because of my poor eyesight — as well as having to do farm work.

At the end of four years of high school, I finished as valedictorian of my class, not a noticeable feat since there were only 18 other students and most of them seldom studied. It was quite an honor as my sister had also graduated with the same award ten years before! For the occasion of graduation, I grew a mustache! I had to give a speech since I was the valedictorian and I was supposed to say something sterling and important? Surprisingly, it went off without a hitch. I gave another student a copy of my speech just in case my memory failed me when I stood on the graduation platform looking out at about 300 parents, friends and other attendees. Fortunately, my memory did not fail me.

Being the valedictorian of my class was important as I received $100.00 for tuition for each semester of my freshman year. At that time, this was the total tuition charged each semester to attend the University of Wisconsin. I had some regrets when leaving high school about what I could have done differently. I did not attend the Junior Prom at my high school because I had no transportation. In high school I was limited to using our old dilapidated farm truck, and with its years of hauling hogs and cattle it was socially unacceptable to use for Junior Prom. I could not dwell on this as my future lay ahead. My entering the University of Wisconsin with its learned professors and a student load of 25,000 was going to be a challenge — especially coming from a high school with a small enrollment and general lack of student facilities.

. . . .

When I went from a small country high school to a major university in 1954, it was quite a change. I was not able to enjoy all the activities offered because I stayed with my parents for both personal and momentary reasons. My mother had suffered a severe stroke in late 1952 and could no longer write or even talk coherently. Consequently, after my parents moved to Madison, Wisconsin, it was necessary for me to be available in the evening hours if she needed any kind of help. My father had retired but found he could not exist on his retirement income and consequently took a job at night working as a janitor. Monetary funds were also a severe problem. I had no income growing up on the family farm, and with my father working at night as a janitor, the only way to attend the university was to stay at home to cut expenses.

I had expressed my disappointment to a fellow student during my freshman year about not being able to participate in many of the university functions because of my personal responsibilities. He replied that I was lucky, he was now trying to move out of the freshman dorm because he was not able to study. He said the majority of the kids never studied and were coming in at 2:00 a.m. in the morning and vomiting their guts out after being sick from drinking beer all night. He said the smell of beer and vomit drifted up and down the corridors of the dorm even through the morning hours.

The University of Wisconsin at Madison has always been described as a party school. Stories abound about the wild parties given at the sorority and fraternity houses. That was not my style and I had no interest in trying to pledge to any fraternity. Even though my mother had severe health problems, she still had one overwhelming quality — she was still an excellent cook. Everything being equal, I would take good cooking over wild parties any time!

I declared geology as my major the first day I registered as a student at the University of Wisconsin. My high school teachers had believed it was a foregone conclusion that I would major in some aspect of farm management. My father had decided, however, that my older brother would work the family dairy farm. That decision was fine with me as I could see no future in farming. I had worked all my young life on that farm and basically had received no compensation for all my efforts. My older sister had felt the same way and had boarded the train for Madison the day after she graduated from high school. Farm work in that period of time was very labor intensive and never seemed to end.

It was a requirement of the Geology Department that every geology major attend a field camp. I took the field camp my sophomore year as I wanted to work summers the next two years. The purpose of the camp was to let geology students become acquainted with the mechanics of geology. It is one thing to attend lectures, and look at rocks, fossils, and minerals in the laboratory, but quite another thing to see these elements in a field environment. The Geology Department offered two types of field courses, a very basic one that let students measure and describe rock units in the field, and a more advanced course giving students the experience of making geologic maps.

For most geology students, this field course is about the only chance they will get to actually work in a field environment. Most geology students work for an oil and gas company that does not believe that field work is very important. Major oil companies in the past have usually required beginning geologists to spend three to six months engaged in some type of field work so that the student can visualize these rock units when they evaluating subsurface electric logs of a drill hole. With the change in oil and gas exploration to deep water exploration and development, as well as the current "fracking" of shale units whose parameters are well known, even those field requirements probably have ended. Field geology has now become a lost art in today's world, but it is still is a valuable tool for exploration when rock units are exposed that can provide good exploration leads. This applies to both mineral and oil and gas exploration.

The field camp held in 1956 was not a field camp in the traditional sense, being conducted more like an expedition to the Mountains of the Moon in Africa. It was going to extend across the country with the first leg of the trip extending from Madison, Wisconsin, to Cloudcroft, New Mexico, for the initial field course. A student could elect to drive his own car, taking three other field camp students along as passengers. There were very few interstate highways in 1956, almost all driving was done on two-lane roads. All of the cars were without seatbelts. My car had a student from Nigeria, West Africa, with us and immediately had a problem. Every time we tried to eat in a restaurant, the African student was waived towards the kitchen. We nixed that immediately and walked out of several restaurants and finally started the practice of buying all our food in grocery stores and making our own meals. It was only on the last leg of our field trip in the northern states that we started eating in restaurants.

On our way to New Mexico, we carried all our camping gear; each one of us was also limited to one duffle bag. We seldom put up our tent; we blew up our air mattresses and slept either on the ground or on a picnic table in a park. The first night we camped near Cloudcroft, one of the students in our car came down with significant pain due to a kidney stone. He basically was out of the field camp for a week. Every day we traveled up and down the road leading to Alamogordo measuring rock sections along the paved road and describing the lithologies. After 10 days we turned in our field work and packed our tents as it was time to move on to the Bridger Range in southern Montana. It was starting to seem that we spent more time traveling than we did actual field work.

The long drive to Montana proved to be very valuable as it was possible to stop at designated spots to see and hear a very interesting discussion of the geology in that area. Our most interesting night on the road was the night we slept on picnic tables behind the police station in Cheyenne, Wyoming. The next night we spent in Thermopolis, which laid claim to have the World's Largest Mineral Hot Springs pool. We actually put our tent that night next to the Hot Springs pool. We then detoured to spend one night in Yellowstone Park and were there for one night. We could have rented a small cabin for $12.00 for one night but the other three students wouldn't spend the money and we slept in our old and very cold tent. We lost considerable sleep that night because the son of Professor Lewis Cline kept throwing firecrackers and cherry bombs at bears who were trying to raid the garbage cans. It was the 4th of July, and it was cold, cold, cold. The next day we made it to the Bridger Range, which is only about 50 miles north of the Wyoming border.

The advance course was spent on mapping the geology of the area. Every day we climbed over 1500 feet or more as we tried to solve the geologic mysteries. After what seemed like an eternity, the marathon was over and we packed up our tents and headed east. A week after returning to Madison, one of the students who had been on the field trip fell off the end of a boat on Lake Mendota and drowned. Somebody by the name of Elvis Presley now seem to hog all the airways. One of the girls on the trip married her driver. It was haying time on the farm and I had to go help pick up 100# hay bales, and then I had to . . .

. . . .

My mother died in August of 1957 when I was between my junior and senior years in college. My father had found my mother on the bathroom floor when he returned home from working until midnight. He immediately called for an ambulance. Word reached me in the early afternoon where I was working in central Wisconsin, and I immediately left for Madison. I found my father sitting outside the hospital room when a nurse came out of the room and said that we could go in to see her. She was conscious enough to recognize me and whispered that she wanted me to recite the Lord's Prayer. I tried the best I could but stumbled over most of the words. The nurse came in and asked us to step outside for a few minutes. As we were starting to sit down in the hallway, the nurse came out and said we had better go back in at once. We did and found she had stopped breathing. She had apparently waited for her younger son to come and bid her goodbye!

. . . .

So it was over, graduation at last in June of 1958. When I finally went across the platform to receive my diploma, it was sort of a disappointment. I was handed an empty folder because the final grades had not been posted. The Assistant Dean didn't even shake my hand but handed me the empty folder and said, "Hurry up; we're behind schedule." Four years and hundreds of hours of study apparently didn't count for much! But still the letters keep coming: "Wouldn't you like to contribute to the University Fund, the Alumni Fund, ..."

Chapter 18

LIFE AFTER DEATH — A SCIENTIFIC EXPLANATION

I tell you, I saw an angel

Every individual who has major surgery has to consider whether he will survive the operation; I was no exception. In my case, open-heart surgery to replace a faulty aortic valve plus three other by-pass procedures would have to be considered major. My choices to consider were very limited. If I DECIDED NOT TO HAVE SURGERY, I would probably have a major — if not fatal — heart attack within several weeks or even days. If I went through with the operation, there would still be a slight chance, five to six percent the surgeon assured me, of my not making it. I decided the odds were in my favor. If I didn't survive the operation, then what?

In the early 1920s, researchers tried to earn a $25,000 prize for proving that a human being had a "soul," i.e., something that would survive after death. They took seven indigent male individuals that were near death and placed them in a special room where they could be closely monitored. These individuals were weighed just before and right after death. Weight loss did occur, about seven grams for each individual. The committee in charge of the experiment declined to grant the $25,000 prize, saying the experiment was not conclusive.

While the research was not considered to be decisive by that committee, more advanced research since that time may indicate a more compelling rationale for the biblical saying, "If you believe in me, you will never die." The whole concept comes down to positive and negative energy always being in absolute balance. The scientists explain it this way:

> **Suppose you were to dig a big hole in the ground and throw all the dirt into a big pile next to the hole you just dug — that pile would represent positive energy. The empty hole is an exact mirror image of the dirt pile and would represent negative energy. To have everything back in balance, the dirt would have to be thrown back in the original hole.**

It is this understanding of how life exists that enables us to determine our future existence. Every human individual has to be completely in balance with respect to positive and negative energy. All life forms, whether human or not, must have this absolute balance of energy. When the death of any individual occurs, the positive energy that was present will disappear, since it is found in the physical part of the body. However, the negative energy still exists, and it is now in an unbalanced state. For existence to continue, it must now combine with another form of positive energy. The negative energy will actually be the blueprint for the continued existence of that individual.

In the example previously given of a hole being dug in the ground, the dirt pile could be removed (positive energy), while the empty hole would still remain. The empty hole in the ground would represent negative energy. Any kind of material could be used to refill the hole, thus restoring the original form that existed. However, depending on the type of material used to fill the hole, it would not exactly resemble the original state of the ground that existed before the hole was dug.

Quantum physics has developed some mathematical equations that indicate different types of dimensions must exist. Any dimension would have to be considered a different energy level in time and space. When an individual dies, his negative energy would immediately try to combine with positive energy from another dimension to restore a complete balance for existence. The individual would now still exist, but in a much different form. Under this concept, the individual will never die, thus confirming the biblical saying, "If you believe in me, you will never die."

Death then, can almost be considered to be self-regulating. Negative energy after an individual's death will always want to combine with the same type of positive energy that had previously existed. The life and belief system by an individual will establish their energy dimensions that eventually will translate over to their future existence in a different dimension after their physical death. All lower life, whether of animal or plant origin, contains the same type of balanced energy. However, when their normal life cycle comes to an end, there is no equivalent positive energy at their lower level for them to combine with. Therefore, their existence comes to an end.

. . . .

In any consideration of the continuation of life after death, it would be of great value to talk or discuss this matter with somebody who had an actual "after life" experience. Fortunately, I did meet somebody, and it turned out to be one of the most interesting experiences of my life. So let me tell you my story. It happened after I had been in the Intensive Care Unit for 14 days, the hospital for four days, and then went to a rehab and nursing home . . .

. . . .

The old minister leaned across the small lunch table in the dining room of a skilled nursing facility and spoke with a loud whisper, "I have seen an angel — I'm not kidding." I met this minister because of my stay in the same facility with him after my open-heart surgery. An aide had wheeled me into the dining room and had placed me at his table. I learned in my many conversations with him that five years ago he had the same heart surgery that I had just endured, so we had some medical experiences in common. One of our conversations at lunch had started quite innocently when I saw this young, very attractive Asian woman wandering around the dining area with a ten-year-old girl following her. When I commented at this, the old minister said, "That's my wife." I apparently looked rather startled because he had previously told me he was 83 years old. Then, as the conversation continued, he told me he became a father when he was 73 years old.

I had reason to doubt his story about an angel, but his story about marrying his young wife seemed bizarre enough that any angel story would probably seem plausible. He had gone to the Philippines and attended a rather unique bazaar where a number of young women would congregate seeking a husband. He found one he liked, and was married and back in the United States within several weeks. Although this whole event seems unreal, there is no future in some of these countries for even the very well-educated women, so foreign husbands are very desirable.

After his wife story, my curiosity about the angel appearance finally led me to ask him more about the angel. His description of an angel was quite interesting. He was quite adamant about what he had seen and he repeated the same story when I asked him about it again. After surgery, he had reached the point in the Intensive Care Unit where he could move around quite freely. He had the feeling one night that somebody else was in the room. Turning over in bed, he stared at a young woman standing at the foot of his bed — blonde, wearing a brown robe with a silver colored sash at her waist. "In addition," he said, "She had some kind of napkin on her head. She looked at me, and I looked at her; then she suddenly disappeared."

I tried to make fun of this and I told him, "Paul, things don't look good for you when even the angels don't want you." However, he was again quite adamant about what he had experienced in the ICU that night. His description of an angel was so entirely different from the literature sometimes found in the Christian churches. Had this minister actually experienced one of the Gatekeepers of the Beyond? In the quietness of the morning hours, I still have reasons to wonder. After ten hours of heart surgery and 14 days in the Intensive Care Unit, how close did I come to seeing her too?

After my return to Denver in the spring of 2011, I wanted to talk to him again. My curiosity about the old minister's angel story could no longer be restrained. Unfortunately, I learned he had died several weeks after I had left the nursing facility in 2010. The strain of being transferred to another facility because of insurance coverage plus his poor health had resulted in his having a major stroke. His illness had been too severe and his angel had come to take him home.

When the old minister had originally told me the description of his angel, he had said that she had some kind of napkin on her head. This description never made any sense to me. When I was reviewing some old 8mm movie tapes, I suddenly realized what he could have meant. One of my relatives had briefly appeared in the movie tapes wearing a very small kerchief on her head. She had folded it into a triangle and tied two ends under her chin. This "napkin" description now made some sense. Another explanation however is also quite possible. Some Catholic women wear a small white lace doily on their head when entering their church. Considering everything, this may be the correct answer.

What I remember most about this old minister was a conversation we had several days before I left the skilled nursing facility. Somehow I felt he had a premonition that once I left, we would never see each other again. In that conversation he slowly voiced his concerns that he had not done enough for society and his church. "I should have written more about the church and its future," he said. "I should have told it like it is. Make sure you write about your life and what it means."

As I looked at that proud old minister hunched over in an old faded nursing gown, I vowed that he was right. I would tell my story too — I would tell it for the ages!

AFTERWORD

When I entered the rehab and nursing center to learn to walk again, I didn't expect any kind of a religious experience. I was glad to have cheated death, as that could have occurred in only a few days, hours or even minutes. When I had almost collapsed while mowing my lawn, and my EKG was found to have changed significantly, I was truly a "walking dead man." I was lucky to have escaped without a major heart attack or any type of heart damage. After I left the nursing home, some people asked me if I really believe that the old minister in my story had actually seen an angel. There is no doubt in my mind that he described what he had seen that night.

Although that event happened five years before I met him, it was still very vivid in his mind when he was describing it. He had originally leaned very close to talk to me at lunch time about another personal event in his life. It was obvious he wanted to confide in me as he seemed to regard me as a kindred spirit. He then brought up the subject of the angel. It was like he had been hiding this happening in his life, not knowing who he could trust with his secret or who would believe him!

So, what about the angel? What would anyone expect an angel to look like? In this case, she looked like a human being, not some creature flying around with wings as has been depicted in some of the Christian literature. The main reason I fully believe him is his description of a "napkin" on her head. It was distinctive enough to be noticed and something that he was unfamiliar with in order for him to want to mention it. Obviously, it didn't occur to him that it was something women may wear in a church setting; something that might indicate the type of environment he was now going to enter. It is the description of the silver colored sash around her waist, however, that I find the most interesting.

There are many scenarios that are suggested by this sash. Some of them are:

1. It could indicate a certain level of responsibility

2. It could indicate a certain district or area

3. It could indicate she was an usher in the church

4. It could indicate she was a guide

5. It could indicate which dimension she was in; i.e., in time and space

More possibilities could be mentioned, but I believe that these would cover the main ones.

My vote is that her sash color indicted what dimension she was in. If one could combine science with Christianity, we may find an interesting possibility. A sociologist, who was quoted in the *Denver Post* newspaper a few years ago, estimated that 50 billion individuals have lived since the dawn of civilization who could be considered "human." The question then becomes, if you are God, would you put Neanderthal Man together with the apostles of Jesus Christ in heaven? Obviously, this doesn't seem logical. It would be better to segregate different cultures in different dimensions that correspond with the way and time that they existed. The bible never describes what "paradise" looks like!

. . . .

If a theologian and a scientific authority in quantum mechanics were placed in a room together and asked about their future existence after death, what do you suppose they would say? Just getting them together in the same room would be quite an accomplishment, because each one is quite satisfied with their concept of reality. The theologian would probably say, "God is everywhere and will save me," and the scientist would ask, "Where is God?" More specifically, we would give them the task of reconciling the concept of Heaven, Paradise, Neanderthal man, and the appearance of the angel at the nursing center, into a plausible story of reality.

Working together, they would probably come to the conclusion that Heaven and Paradise are the same — but different. "Heaven," they would say, "is the utmost center of Paradise." Going to Paradise after death, is not the same thing as going to Heaven. Heaven is where the most absolute authority would exist. Paradise would be considered a dimension in which Heaven would be found!

The appearance of the angel in the story of the minister in the nursing home is not difficult to reconcile with quantum mechanics when one considers the quantum leap theory. In an example often cited, an electron is assumed to whirl around a nucleus of an atom in a certain pattern that is fixed, always staying a certain distance from the nucleus. It, however, has the ability to leap to another path of equal status around the same nucleus as it continues on its journey. If we now change our concept from pathway to that of a dimension, our understanding of reality becomes much clearer. The angel, for whatever reason, had the ability to change dimensions. She ended up at the nursing home because…?

Neanderthal man would also occupy a dimension, probably one that would lie further away from heaven. He would not occupy the same level or dimension as the Apostles of Jesus Christ. These different dimensions are probably what the Apostle Paul was thinking about when he cited the existence of other "Principalities." It is for this reason that Jesus Christ is recorded in the bible to have said, "Father, into your hands I command my spirit." — he didn't want to end up in the wrong dimension. In this whole scenario, where is Hell. Hell would have to be considered the lowest dimension, one that is earthbound. Anything connected with planet earth, at whatever level, will cease to exist when the whole solar system is pulled into the sun as it collapses when it runs out of fuel. The end will not be very pleasant.

Chapter 19

TALKING SNAKES — A CHRISTIAN DILEMMA

It was becoming quite obvious that the young woman was becoming more and more agitated! While I was standing on a street corner on a very hot, humid Sunday afternoon in Tulsa, Oklahoma, I had noticed the street preacher approaching her holding his Christian tracts. Their conversation apparently had started quite pleasant enough. Suddenly, the young woman started raising her voice and waving her arms in the air. As the argument grew more heated, her voice cut through the din of street noise and road traffic. "You want to know why I'm not a Christian," she yelled, apparently not giving heed to any of the nearby pedestrians. "I will tell you why I'm not a Christian if you really want to know. <u>I'm not a Christian because I don't believe in talking snakes</u>."

This street argument came to my mind as I sat in my church the next Christmas Eve. The old familiar hymns of the church and of my boyhood had echoed throughout the church that evening, giving me an overwhelming feeling of peace and contentment. Yet for some reason, I could not forget that street argument months before. As the service ended, and the parishioners filed out, I couldn't help but notice the almost universal appearance of gray hair and bald heads. *Where are all the young people with families?* I thought. There must be somebody under the age of forty? As I sat there in the quiet of the evening, the scene was starting to accentuate an overwhelming feeling that I was beginning to have — *the Christian Church was its own worst enemy!*

Still sitting there, I also remembered one time I had gone to a men's Good Friday breakfast at a local church. As I entered the church door, I was confronted by two older, well dressed men waving bibles in the air, obviously self-proclaimed missionaries of the Christian Church. They told me that the bible was the absolute word of God and could not be questioned or challenged! They said I had to confess the Christian faith that morning to save me from eternal damnation!

As I stood there listening to them, I finally said, "If all you say is true, why are there several hundred protestant denominations? Also tell me something, who did Cain marry after he killed Abel in the biblical story? Only Adam, Eve, Cain and Abel were supposed to have existed? Also, which of the two creation stories given in Chapters One and Two in Genesis is the correct one?" They just stood there, glaring at me. Hearing nothing, I handed them back their literature and entered the interior door of the church. I should have taken the time to debate them further, I thought, but frankly I was hungry. It was obvious from their arguments that they had fallen into the fallacy of trying to use the bible to try and explain everything in existence.

The information age we now live in is partly responsible for the decline of church attendance and membership. There has always been a divide between scientific research and understanding and Christian theology, but it has become more pronounced in recent years. Each general group retreats into its own philosophy. A theologian doesn't take science courses, and a scientist doesn't study theology. In the early 1930s, Billy Graham learned that a colleague had taken a course in geology, and was horrified! "Don't get too scientific," he said, "just preach Jesus Christ crucified!"

Many theologians will never quite understand why science cannot accept everything the way it is expressed in the bible. This is mainly because they don't understand the way science functions. It is one thing to say you don't understand how a light bulb works or how a car engine can run. It is quite another thing when you are considering your future life existence. <u>Science builds on facts. When the underlying factual basis has some important discrepancies, doubt will be created because one cannot now come to a logical and understandable conclusion; believability becomes a problem.</u>

Unfortunately, science and Christian theology will never meet an acceptable understanding with each other. Science is always searching for the ultimate truth and will readily and quickly change interpretations of its own belief systems if necessary. Christian theology, however, has a huge problem. It is stuck with its dogma! It cannot change, because this would undermine its belief doctrines, as well as lead to an attack on the bible itself. A current example is a very liberal group of theologians who studied the biblical sayings of Jesus Christ as given in the bible. In their opinion, about 84% of what he was reported to have preached was not his sayings. How they reached their conclusions is unclear because none of them were there 2000 years ago — while others were.

The Christian bible reached its current form somewhere around 325 A.D. when the church fathers finally decided what writings should be placed in it. One example of their decision would be the Book of Esther. This was placed in the bible even though many modern day scholars debate its theological underpinnings. The Gospel of Thomas was not included because they considered it was written too long after the biblical history of events. The early writings of the Jewish scribes occurred after 587 B.C. when they were in exile in Babylon. They needed a holy book because they were not able to worship in their temple in Jerusalem. Prior to this date, their knowledge was passed on by oral traditions.

The opening chapters of Genesis, the first book of the Old Testament, is, and will always be, a major headache for the Christian Church. Most average young adults without a Christian background and with a curiosity about the bible, will pick up the bible and start reading Genesis. At that point their doubts begin. They wonder what is the "world" that the scribe is writing about. What is the "world" to a young religious scribe sitting in a goat-hide tent along the Euphrates River 4,000 years ago. For example, has he ever been more than 20 to 30 miles from his tent? Why doesn't he mention or write about the great lands outside of where he is now living — like North and South America, Australia, Iceland, Greenland, Europe, and Asia. Obviously, he seems to have some limitations about what he knows.

The questioning of youth continues. If God exists forever, why does he have different days of the week. How about Noah's Ark. It was supposed to be made of wood logs and reeds — not quite as big as the Titanic. That ship was built of the best steel construction and technology available, thousands of years after the Ark, but it sank when it hit some ice. The Ark would have been held together with vines — how would it stand up in even a minor storm? If the whole world was covered with water, where did the water come from? — and where did it go? If water were to cover the entire earth, even above highest mountain peak, wouldn't some of the animals have died from lack of oxygen? Also, if the earth rotates every 24 hours, does heaven also rotate? If it does rotate, what is up then becomes down, and does down become Hell — or is it still up?

What about the age of the earth — do the theologians study earth history? Wasn't it too many years ago that the Christian Church burned hundreds of people at the stake because they wouldn't believe that the sun went around the earth — and didn't they also say that the earth was flat? Why are certain animals like kangaroos restricted only to Australia if evolution doesn't exist. Why do people before the flood live over 600 years, but only a little over a 100 years after the flood? Also, which zoo has some talking snakes?

These types of questions will continue to multiply until the Christian Church faces these questions directly. Why, for example, does the church want to deny evolution? The answer is simple enough, to do so would then imply that the church believes that man descended from a monkey.

This would be contrary to the biblical belief that man was created by God. The question, however, remains for both science and theology as to when did man become MAN?

Science would like to believe that man started to exist three to four million years ago, while Christian theology would like it to be around four to six thousand years ago. Actually, one could consider both belief systems to be correct. Science has its proof in carbon dating for its age determination. Christianity can lay claim to Abraham. Abraham is the first MAN to state his belief in an eternal God and to follow his perceived teachings. Science is considering only the physical nature of man, while Christianity is looking at SPIRITUAL MAN. Each side is considering only what is important to their basic understanding.

When one studies the bible, it is readily apparent that a line of demarcation can be drawn across biblical history following the ancient story of the world-wide flood. Before the flood, snakes can talk, God makes a woman out of a rib, men marry women who don't exist, a big wooden boat is made that rivals what modern technology can make, and God, who lives forever, takes a day off after making the world. It is readily apparent that the Jewish scribe has borrowed the traditions and mores of previous cultures to try and explain his own. This borrowing of prehistory creates a confusion that undermines the authenticity of the bible itself. Beginning with Abraham, there can be no doubt of Jewish history or its origins, and eventually of Christianity.

The story of the Ark and the biblical flood will forever be found in every Christian school classroom. Every conservative Christian theologian will always declare it to be the absolute truth since it is found in Genesis. Where this story originally came from is very interesting. The ancient world believed that the sky and clouds represented another sea held up by four big pillars. They also believed that they were standing on an island that floated on another sea. Thus, it was very easy to believe that enough water could be obtained to cover the entire "earth" and also get rid of it (Genesis 8:2). The world-wide flood story was a staple of almost all cultures in the ancient world as they tried to explain their origins.

In ancient history, time was also measured sometimes in "moon years." The given age of an individual in the bible would be inflated in some cases depending on the source the scribe is quoting. Noah's age of 950 years when he died becomes 79 years when divided by the 12 moon cycles in a year. This disparity in age is another indication that the Jewish scribes borrowed from prehistory to describe his own. The conservative theologian will try to discredit "moon years" because it would destroy his concept as the exact date that God created the world.

It is also unfortunate that it was within the magnitude of the ancient scribes to enlarge and make some event bigger than it was because they wanted to show its importance. In the story of the Exodus, the biblical scribe enlarges the number of Jewish people fleeing Egypt. The number given is about 600,000 men, women, and children, plus their animals and household goods. This total number did not include the Levites who were given the responsibility of the priesthood. Given the recent Katrina tragedy in New Orleans, the U.S. Government and other agencies could not even evacuate 50,000 people out of the city in three days with one of the most modern transportation systems known to mankind.

The Christian church has gone to great lengths to make the bible into a much more understandable and readable book of theology. Why the church continues to neglect giving the reader a greater understanding of the cultures from which the book evolved is very unclear. It is quite clear that the bible has to be evaluated according to the belief systems of the ancient scribe. It is not enough to simply "clean up" the language in another revision of the bible to make it more readable. It must make some aspects more understandable as well. The Christian Church cannot continue to allow some unclear written aspects of ancient history to undercut and erode the wisdom and truth of the bible.

However, nothing will ever change. The Christian Church will always pretend to know all the answers, and will never allow questioning of anything written in the bible. When questioning does occur, the standard answer will be, "God can do anything." Theoretically speaking, this gets the church off the hook without having to explain anything. Maybe Billy Graham did have the right answer, "Don't get too scientific." However, the church will continue to lose membership and the youth will go elsewhere. In the not too distant future, people will be riding the light rail and will point out to their children a dilapidated, rundown, and abandoned building and say, "See, that's what they used to call a Christian Church."

AFTERWORD

When the average Christian picks up the Holy Bible, he doesn't realize that he is looking at one of the oddities in history. All the individuals mentioned in the Old Testament are either members of the Jewish faith or are gentiles! Indeed, the Old Testament could almost be called a "Jewish Bible." The Christian Church is unique among religions in that it reaches out and grabs another religion and uses it as part of its own theology. When a religion does this, it causes immense problems in the understanding of its beliefs because it is stuck with various tenets of the usurped religion. The Old Testament shows that the Jewish faith is based on The Law, while the New Testament of the Christian Church is based on Faith. Reconciling the two becomes a problem. A lot of misunderstanding in theology interpretation could have been avoided if the early church had regarded the Old Testament as a historical record rather than give it the same theological status as the New Testament!

The early leaders of the Christian Church sought to prove that it was now really the righteous religion by accepting everything in the Old Testament as true and factual in order to prove and verify their own beliefs. On an average Sunday morning in the Protestant churches, probably 70% or more of the sermons are based on Jewish scripture of the Old Testament. Although the Old Testament is based on the Jewish understanding of their God given law, it comes into conflict in understanding with the Faith concept given in the New Testament. Many authors have described the God of the Old Testament as being warlike and vengeful. The God of the New Testament is depicted as loving, kind, and forgiving. The comparison between these two extremes is very striking!

The fall of Jerusalem in 70 A.D. is indicative of the challenges facing the Christian Church of today and why it continues to lose influence in the everyday world. Josephus Flavius, an early Jewish historian, records the Jewish defenders constantly fought among themselves and could never unite in a successful effort to break the Roman siege in A.D. 66–70. Under a united army with an adequate military commander, the combined Jewish army could probably have broken the Roman siege and changed world history. With a large Jewish unifying force in the Middle East throughout history, doubt has to be created that the Moslem faith would have become as dominant as it is today. The fragmentation of the Christian Church in today's world is not giving the Church enough credence to be an effective force in our society.

The constant diving into the Old Testament by ministers for sermon material is counter-productive to our understanding of Christian beliefs. I have been amazed when esteemed Christian leaders, like Billy Graham, give sermons about Noah and the Ark. When archeologists found the ancient Assyrian capital city of Ninevah, in what is now northern Iraq, they discovered a complete library on cuneiform clay tablets from the 7th century B.C. One of the stories that emerged was an old Sumerian story called the Epic of Gilgamesh, a story that parallels the Noah and Ark story. These clay tablets were quoting older flood stories several thousand years before that date. Similar flood stories have emerged from many different areas of the Middle East, all pre-dating the Noah and the Ark story. Nothing can be gained by Christian ministers or theologians by quoting old Jewish stories about a universal flood.

A far better story to be placed in Sunday School classrooms concerns the three astrologers or "wise men" who came to Bethlehem with gifts of frankincense, myrrh, and gold. Frankincense is a fragrant gum resin obtained from trees in East Africa and is an important incense resin. Myrrh is a reddish-brown aromatic gum resin that has a bitter taste, but apparently was mixed with labdanum which is a dark oleoresin used in biblical times for making perfumes. This combination of gifts seems rather odd to be given to a poor family. Granted, these gifts were for a king; but, perfume, for a king? Could it be that two of these three gifts, incense and perfume, were actually meant for the bazaars of Egypt?

The story of a child born in a stable is at the very heart of Christianity. A very valid reason for this birth to take place in a stable at that time in history is that it provided ready access for transportation to Egypt. Joseph had to seek instant exit from Bethlehem since that city was only eight miles from King Herod. No one can cross 700 miles of sun drenched desert filled with bandits and sand storms on just a single donkey and no water or supplies. A camel caravan was the only way to go and travel could take up to a month. Joseph would know what transportation was immediately available because of the circumstances he found himself in. Because of the gifts given by the wise men, he had the resources available for the trip to and from Egypt as well as the stay in Egypt.

Now think of the story that can be told! A young family watches a troop of soldiers loyal to King Herod walking into Bethlehem as the camel caravan slowly winds itself out of the city on its way to Egypt. The three gifts given by the "wise men" were now going to be used for the safety and welfare of this young family in a foreign land. Now the New Testament can eventually turn to the history of the Old Testament — Out of Egypt I have called my Son!

. . . .

It is probably disheartening to many Christians to realize that most of the Old Testament <u>would not</u> be admitted into a Court of Law in this country because it would be considered "hearsay" — only the New Testament bears witness to the GOOD NEWS. Reliance on myths by Christian theologians related to the early Jewish writings have seriously damaged the Bible in its credibility among scientists and well-educated people. The Christian Church will never change its interpretations of history and will cause more harm than good for those individuals trying to seek the Truth and Wisdom of God!

There is an area where science and Christianity do reach an absolute agreement, and that is the concept of HELL. Both sides come to the same conclusions, but by different methods. Science has determined that the sun will exhaust its fuel in about five billion years and will become a red giant that will extend out to the orbit of the earth, basically destroying the earth and everything on it. Eventually, the sun will collapse inward pulling the entire solar system with it. That total collapse into a fiery end will even make the Christian version of hell seem tame by comparison. Whether you're a scientist or a theologian, you don't want to be on planet earth when that finally happens.

EPILOGUE

Is Washington, D.C. Ready for a Big Earthquake?

The recent earthquake damage to the Washington Monument should have been a wakeup call to the U.S. Congress to start providing additional funding for more geologic research in the eastern part of the United States. For some reason, there seems to be a perception that the research activities of the U.S. Geological Survey are to be directed mainly to the Western States as that is where the federal lands are located. However, 60% or more of the American population lies along the eastern seaboard where earthquake damage could be significant. Only when the White House is sitting on top of the Congressional Office Building due to an earthquake, would there be a perception in Congress that information is probably needed about the geologic structure of the area. By that time, it will be too late.

Earthquake damage in the East has not been significant in the recent past. The earthquake that damaged the Washington Monument was about a magnitude of 5.8; however, the damage it caused was fairly extensive. When work started on restoration, the needed repair of the building was found to be more than had been originally anticipated. The amount of damage is indicative of what could happen if an earthquake of much larger magnitude occurred in the Washington, D.C. area.

What is needed in the very eastern part of the United States is a concerted effort to implement a geologic mapping program that resembles the mapping program in the Western States in the 1960s and 1970s. That mapping program produced geologic maps that covered 8,000 miles of territory. These maps provided the basis for an accurate analysis of the type of geologic structure occurring throughout those areas.

What currently exists along the eastern seaboard is a mixed bag of different maps, some produced by the U.S. Geological Survey, others by state surveys and university studies. In many cases, the maps are not of sufficient scale or detail to be definitive for adequate analysis. <u>Funding must not</u> be given over to state surveys and universities as has been done occasionally in the past. A regional detailed framework can only be completed by the U.S. Geological Survey. This would prevent problems of different analyses occurring at artificial boundaries, i.e., state boundaries, etc.

The cost of the mapping program is relatively inexpensive for the type of information that will be obtained. Approximately 10 to 12 million dollars a year would be a reasonable estimate. The total cost of over 100 million dollars is minor compared to the havoc that a major quake of 6.8 or more would produce. At first glance, the program might seem expensive, but compared to other government programs it is not. Consider the cost to the U.S. Government to repair and refit one Army Abrams battle tank. Currently, the military has about 8,000 Abrams tanks in inventory. At any one time, about 700 tanks need repair of some type. <u>The cost to repair and refit one tank is about one million dollars</u>. If the repair of ten tanks were delayed every year in the repair cycle, it would fund the cost of a mapping program that could save thousands of lives.

Any program designed to determine if dangerous earthquake potential exists on the East coast must be science driven. The debacle created by the politically driven CRO/CDP maps described in the headlines of the *Denver Post* on April 29, 1979, are indicative of the problems that are caused by political interference in the scientific process. See the article by the *Denver Post* in the Appendix.

BIBLIOGRAPHY

1. Schroeder, M.L. **Arenaceous Foraminifera from the Shawnee Group of the Upper Pennsylvania in Southeastern Kansas**. Master's thesis, University of Kansas, 1961.

2. Jobin, D.A. & Schroeder, M.L. **Geology of the Conant Valley quadrangle, Bonneville County, Idaho**. U.S. Geological Survey Mineral Investigations Field Studies Map MF-277, 1964.

3. Jobin, D.A. & Schroeder, M.L. **Geology of the Irwin quadrangle, Bonneville County, Idaho**. U.S. Geological Survey Mineral Investigation Field Studies Map MF-287, 1964

4. Albee, H.F., Jobin, D.A. & Schroeder, M.L. **Northwesterly extension of the Darby Thrust in the Snake River Range, Wyoming and Idaho**. U.S. Geological Survey Prof. Paper 575-D; p. D1-D3, 1967.

5. Pampeyan, E.H., Schroeder, M.L., Schell, E.M., & Cressman, E.R. **Geologic map of the Driggs 15-minute quadrangle, Bonneville and Teton Counties, Idaho and Teton County, Wyoming**. U.S. Geological Survey Mineral Investigation Field Studies Map MF-300, 1967.

6. Schroeder, M.L. **Lower Triassic foraminifera from the Thaynes Formation in southeastern Idaho and western Wyoming**. *Micropaleontology*, vol. 14, no. 1, pp. 73–82, Jan., 1968.

7. Schroeder, M.L. **Geologic map of the Teton Pass quadrangle, Teton County, Wyoming.** U.S. Geological Survey Quadrangle Map GQ-793, 1969.

8. Schroeder, M.L. **Geologic map of the Rendezvous Peak quadrangle, Teton County, Wyoming.** U.S. Geological Survey Quadrangle Map GQ-980, 1972.

9. Schroeder, M.L. **Preliminary geological map of the Clause Peak quadrangle, Lincoln, Sublette, and Teton Counties, Wyoming: Map with cross sections and table of chemical analyses of phosphatic rock.** U.S. Geological Survey Open-file report no. 71-250.

10. Schroeder, M.L. **Geologic map of the Clause Peak quadrangle, Lincoln, Sublette, and Teton Counties, Wyoming.** U.S. Geological Survey Quadrangle Map GQ-1092, 1973.

11. Schroeder, M.L. **Preliminary geologic map of the Camp Davis quadrangle, Teton County, Wyoming,** U.S. Geological Survey Open-file report no. 73-250.

12. Schroeder, M.L. **Geologic map of the Camp Davis quadrangle, Teton County, Wyoming.** U.S. Geological Survey Quadrangle Map GQ-1160, 1974.

13. Schroeder, M.L. **Geologic map of the Bull Creek quadrangle, Teton and Sublette Counties, Wyoming**. U.S. Geological Survey Quadrangle Map GQ-1300, 1976.

14. Schroeder, M.L. **Preliminary geologic map and coal resources of the east half of the Guild Hollow quadrangle, Uinta County, Wyoming**. U.S. Geological Survey Open-file report OF 77-427, 1977.

15. Schroeder, M.L. **Preliminary geologic map and coal resources of the Ragan quadrangle, Uinta County, Wyoming**. U.S. Geological Survey Quadrangle Map C-85.

16. Schroeder, M.L. & M. Dronyk; **Lithologic and geophysical logs of 30 coal test holes drilled in the Hanna Basin coal field, Carbon County, Wyoming**. U.S. Geological Survey Open-file report OF 78-657, 1978.

17. Schroeder, M.L. **Geophysical logs of 15 coal test holes drilled in the Kemmerer coal field, Uinta County, Wyoming**. U.S. Geological Survey Open-file report OF 78-658, 1978.

18. Schroeder, M.L. **Lithographic and geophysical logs of 16 coal test holes drilled in the Rock Springs coal field, Sweetwater County, Wyoming**. U.S. Geological Survey Open-file report OF 78-659, 1978.

19. Schroeder, M.L. **Preliminary geologic map of the Pickle Pass quadrangle, Teton County, Wyoming**. U.S. Geological Survey Open-file report of 78-1630, 1979.

20. Schroeder, M.L. **Preliminary geologic map and coal sections of the Sulphur Creek quadrangle, Uinta County, Wyoming.** U.S. Geological Survey Open-file report OF 79-1631, 1979.

21. Schroeder, M.L. **Geologic map of the Pickle Pass quadrangle, Lincoln County, Wyoming.** U.S. Geological Survey Quadrangle Map GQ-1630.

22. Schroeder, M.L. **Preliminary geologic map of the west half of the Bridger quadrangle, Uinta County, Wyoming.** U.S. Geological Survey Open-file report OF 79-1632, 1979.

23. Schroeder, M.L. & Lunceford, R.A. **Preliminary geologic map and coal sections of the Cumberland Gap quadrangle, Lincoln and Uinta Counties, Wyoming.** U.S. Geological Survey Open-file report OF 79-1633, 1979.

24. Schroeder, M.L. **Geophysical logs of 5 coal test holes drilled in the Kemmerer coal field, Uinta County, Wyoming.** U.S. Geological Survey Open-File report OF 80-1026, 1979.

25. Schroeder, M.L., Albee, H.F., & Lunceford, R.A. **Geologic map of the Pine Creek quadrangle, Lincoln and Teton Counties, Wyoming.** U.S. Geological Survey Quadrangle map GQ-1549, 1981.

26. Schroeder, M.L. **Geologic map of the Deer Creek quadrangle, Lincoln and Teton Counties, Wyoming.** U.S. Geological Survey Quadrangle Map GQ-1551, 1981.

27. Schroeder, M.L. **Geologic map of the Bailey Lake quadrangle, Teton County, Wyoming.** U.S. Geological Survey Open-file report OF 81-715

28. Schroeder, M.L. **Preliminary geologic map and coal sections of the Elkol SW quadrangle, Lincoln and Uinta Counties, Wyoming.** U.S. Geological Survey Open-file report OF 81-716.

29. Schroeder, M.L. **Preliminary geologic map and coal sections of the Meadow Draw quadrangle, Lincoln and Uinta Counties, Wyoming.** (Intended for publication in the coal series).

30. Schroeder, M.L. **Geologic map of the Bailey Lake quadrangle, Lincoln and Teton Counties, Wyoming.** U.S. Geological Survey quadrangle map GQ-1608, 1987.

GENERAL GEOLOGIC REFERENCES

Armstrong, F.C. and Oriel, S.S., 1965. **Tectonic development of Idaho-Wyoming thrust belt**. *Amer. Assoc. Petrol. Geol. Bull.*, v. 49, no. 11, p.1847–1866.

Blasdell, E., 1969. **Geology of the Granite Creek area, Gros Ventre Mountains, Wyoming: Master's Thesis**. Univ. of Michigan, Dept. of Geology and Mineralogy, 49 p.

Dorr, J.A., JR., 1952. **Early Cenozoic stratigraphy and vertebrate paleontology of the Hoback Basin, Wyoming**. *Geol. Soc. Amer. Bull.*, v.63, p. 59–94.

Dorr, J.A., JR., 1956. **Post-Cretaceous geologic history of the Hoback Basin area, central western Wyoming:** *Wyoming Geol. Assoc. Guidebook.* 11th Annual Field Conference, p. 99-1080

Eardley, A.J., Horberg, L., Nelson, V.E., and Church, V., 1944. **Hoback Gros Ventre-Teton Field Conference map**. Conference arranged by the staff of Camp Davis, the University of Michigan Rocky Mountain Field Station, privately printed.

Eardley, A.J., 1944. **Geologic map (of Camp Davis area)** prepared for Hoback-Gros Ventre — Teton field conference: Univ. Mich.

Foster, Helen, 1943. **Structure and time relations of a portion of the Hoback Range and Basin**. Dept. of Geology, The Univ. of Michigan, Ann Arbor, M.S. dissertation, 47 p.

Froidevaux, C.M., 1968, **Geology of the Hoback Peak area in the overthrust belt, Lincoln and Sublette Counties, Wyoming**. Dept of Geology, Univ. of Wyoming, M.S. dissertation, 126 p.

Guidebook to the Energy Resources of the Piceance Creek Basin, Colorado. RMAG 25 Annual Field Conference, 1974.

Hayden, E.N. **History of Jackson Hole**. Univ. of Wyoming, in Jackson Hole Field Conference Guidebook, 1956, Wyoming Geologic Survey.

Horberg, L., 1938. **The structural geology and physiography of the Teton Pass area, Wyoming**. Augustana Library Publication, no. 16, p. i-x and 86.

Horberg, L., Nelson, B., and Church, V., 1949. **Structural trends in central western Wyoming**. *Geol. Soc. Amer. Bull.*, v. 60, p. 183–216.

Keefer, W.R., 1964. **Preliminary report on the structure of the southeast Gros Ventre mountains, Wyoming**. U.S. Geological Survey Prof. Paper 501-D, p. D22–D27.

Love, C.M., 1968. **The geology surrounding the headwaters of Nowlin, Flat, and Granite Creks, Gros Ventre Range, Teton County, Wyoming**. Masters Thesis, Montana State Univ., Bozeman, 106 p.

Love, J.D., 1956b. **Summary of geologic history of Teton County, Wyoming, during late Cretaceous, Tertiary and Quaternary times**. *Wyoming Geol. Assoc. Guidebook*, 11th Annual Field Conference, p. 140–150.

Love, J.D., and Albee, H.F., 1972. **Geologic map of the Jackson Quadrangle, Teton County, Wyoming (1:24,000)**. U.S. Geol. Survey Map I-769-A.

Love, J.D., and Reed, J.C., Jr., 1968. **Creation of the Teton landscape**. *Grand Teton Natural History Assoc.*, 120 p.

Love, J.D., Reed, J.C., Christiansen, R.L., and Stacey, J.R., 1973. **Geologic block diagram and tectonic history of the Teton Region, Wyoming-Idaho**. U.S. Geol. Survey, Misc. Geol. Invest., Map I-730.

Mountjoy, E.W., 1966. **Time of thrusting in Idaho-Wyoming thrust belt — Discussion**. *Amer. Assoc. Petrol. Geol. Bull.*, V. 50, no. 12, p. 2612–2614.

Oriel, S.S., and Armstrong, F.C., 1966. **Time of thrusting in Idaho-Wyoming thrust belt — Reply**. *Amer. Assoc. Petrol. Geol. Bull.*, v. 50, no. 12, p. 2614–2621.

Ross, A.R., and St. John, J.W., 1960. **Geology of the northern Wyoming Range, Wyoming**. *Wyoming Geol. Assoc. Guidebook*, 15th Ann. Field Conf., Overthrust Belt of Southwestern Wyoming and Adjacent Areas, 1960, p.45–56.

Royse, F., Jr., Warner, M.A., and Reese, D.l., 1975. **Thrust Belt structural geometry and related stratigraphic problems, Wyoming-Idaho-northern Utah**. Rocky Mtn. Assoc. Geol., Sym. On Deep Drilling Frontiers of the Central Rocky Mountains, p. 41–54.

Rubey, W.W., and Hubbert, M.k., 1959. **Role of fluid pressure in mechanics of overthrust faulting, II. Overthrust belt in geosynclinal area of western Wyoming in light of fluid-pressure hypothesis**. *Geol. Soc. Amer. Bull.*, v. 70, p. 167–205.

Sehnke, E.D., 1969. **Gravitational gliding structures, Horse Creek Area, Teton County, Wyoming**. Unpublished Masters paper, The Univ. of Michigan, Department of Geology and Mineralogy, Ann Arbor, 41 p.

Thies, B.P., 1974. **Structural configuration of Horse Creek Canyon, Hoback Range, Wyoming**. Masters Thesis, The Univ. of Michigan, Dept. of Geology and Mineralogy, 18p.

Wanless, H.R., Belknap, R.L., and Foster, Helen, 1955. **Paleozoic and Mesozoic rocks of Gros Ventre, Teton, Hoback and Snake River ranges, Wyoming**. Geol. Soc. Amer., Mem. 63, 90 p., 13 pls.

NON-GEOLOGIC REFERENCES

Frady, M. **Billy Graham, A Parable of American Righteousness**. 1979, Little, Brown and Company, Boston, 542 p.

Harris, Barton. **John Colter and His Years in the West**. 1952, originally printed by Scriber.

Hawking, S.W. **The Grand Design**. 2010, Bantam Books, 198 pp.

Hawking, S.W. **The Universe in a Nutshell**. 2001, Bantam Books, 216 pp.

Johanson, D. and L. **Ancestors — In Search of Human Origins**. 1994, Villard Books, New York, 39 p.

Master Study Bible, New American Standard. Holman Bible Publishers, Nashville, 2384 p.

Morgan, Dale, **Jedediah Smith and the Opening of the West**. 1983, Bison Books.

Schwinger, John, **Einstein's Legacy (The Unity of Space and Time)**. 1986, Scientific American Books.

Thiede, C.P. **The Dead Sea Scrolls**. 2000, Lion Publishing, London, 256 p.

Thomsen, B.M., editor. **The Man in the Arena, the selected writings of Theodore Roosevelt**. 384 p.

Vestal, Stanley, Jim Bridger. **Mountain Men**. 1946, Univ. of Nebraska Lincoln.

Will, C.M. **Was Einstein Right? Putting General Relativity to the Test**. 1986, Basic Books, N.Y., 274 p.

Wilson, Ian. **Before the flood**. 2011, St. Martin's Press New York.

APPENDIX

$20 Million Coal-Lode Maps 'Useless'

by Joseph Seldner, *Denver Post* Staff Writer
Reprinted by permission of the *Denver Post*

Work performed in a $20 million program of the U.S. Geological Survey to map the nation's federally owned coal deposits is virtually worthless, several geologists claim.

Poor planning, bad management and inaccurate work have resulted in a multimillion-dollar set of unreliable or useless maps, the critics say.

Much of the criticism comes from within the USGS, which contracted the mapping work to private companies. Outside geologists who have seen the maps also have attacked the quality of work.

One top-level USGS official conceded to his colleagues recently that the agency will have to settle for a finished product that is "something better than crap."

A geology professor at the Colorado School of Mines, who was retained two weeks ago by the *Denver Post* for an independent review of some of the work, concluded that the program has been "an incredible waste of money."

THE MASSIVE program, authorized by Congress, was intended to map all federally owned coal in the United States, the vast majority of which is in the Rocky Mountain West.

Documents and tape recordings obtained by The Post, and interviews with persons involved directly or indirectly with the project, show that almost from the start of the program in 1975 there has been concern over the need for the mapping and the quality of maps produced to date.

About $10 million already has been spent on the project, with at least $10 million more needed to complete it.

The program was designed to map the coal deposits in 1,400 "quadrangles" of land, each covering 35,000 acres. The cost of mapping each quadrangle is about $10,000.

MANY OF THE AREAS contain only a small amount of federal coal, and those maps show large spaces of blank area.

"They're about the stupidest-looking things I ever saw," said Gary Glass, deputy director of the Wyoming State Geological Survey and a recognized coal expert.

The maps don't show privately owned coal formations that are adjacent to federally owned coal deposits, leading one energy-company official to comment in a letter to USGS that the maps are about as useful as a road map of Colorado that charted only roads on federal land.

In his letter, Gary Myers, chief geologist for the Energy Fuels Corp., asked for a refund on the money he paid for a map of the Wolf Mountain quadrangle in northwest Colorado.

Complaints about the maps include cases of inaccurate work. In one case, maps produced by Texas Instruments Inc. show a 15-foot bed of coal abruptly butting against an area containing less than one foot of coal.

BOB HAMILTON, the School of Mines assistant professor of geology who served as a consultant for The Post, described such a formation as "geologically impossible."

Geologists told The Post that such mistakes undermine the credibility of all the maps.

One memorandum from a USGS geologist has warned of the possibility of lawsuits filed by private companies relying on maps which show coal beds in the wrong legal locations.

Another government geologist predicted in a letter to the director of USGS, "When various companies spend thousands of dollars for these maps and find out they are for most purposes unusable, there are liable to be many repercussions."

Jerry Davis, who holds a doctorate in geology, worked on the mapping program for two years before resigning in disgust over mismanagement and the dubious value of the maps.

HE SAID THE program is so bad that to make it worthwhile, "they'd almost have to do everything over again, and that would cost another $10 million."

Don Kash, chief of the USGS division in charge of the project, flew to Denver from USGS headquarters in Reston, VA., early this month and met with unhappy geologists in the agency's regional office here.

In his remarks, he told staff members who suggest that the program be reconsidered, "I haven't got guts enough to throw away $10 million." He indicated the project would be continued.

At one point in the April 2 meeting, however, he cautioned that "we're going to have our problems delivered on the front page of the newspapers one of these days if we don't solve our problems." The project was created to provide maps for the Bureau of Land Management in land-use planning and ultimately to aid in leasing federal land for coal mining.

BUT INTERNAL documents at USGS and comments from persons familiar with the maps indicate that so little information is contained on many of them that they are worthless for land-use planning.

Glass, who said he has reviewed "dozens" of the maps, said they are of no use for land-use planning. The project "ignores all resources on state and private land in any of the areas," he said. "I don't see how you can plan ignoring that."

Davis, who resigned from the program last November, was assigned to review and approve maps produced by the private contractors hired in 1977 after USGS decided it would take too long for its own people to produce them.

Davis told The Post that the project would have been much more efficient if it had focused on coal deposits that can be mined now or in the near future.

"THERE'S A LOT of coal out there and it doesn't mean you have to map areas now that you're going to use in 200 years," Davis said.

Under terms of the contracts, all maps must be made by compiling existing information — no field inspections or on-site evaluations are allowed.

As one contractor explained it, "We're not allowed to leave our office to do the maps."

Specific points of complaint in the program include:

- One map of a quadrangle in Moffat County, Colo., that contained no coal except for two small patches on state-owned land. The mapping contract by the AAA Engineering and Drafting Co. of Salt Lake City (as have all the program's contracts) specified that only federally owned coal be mapped.

- A report from Texas Instruments Inc. that one quadrangle the company was asked to map in Wyoming contained "no coal reserves beneath federal land with the quadrangle."

- Internal memos from USGS staff experts pointing out that maps already exist showing most of the coal information and urging that the contracts be revised or suspended.

- Indications that much of the map work was done by copying information directly from documents drawn up earlier this century and known to be inaccurate.

SEVERAL USGS staffers have warned that the program will be a blot on the agency's integrity and could damage seriously its reputation with private industry.

Because of the poor quality of many of the maps turned in by contractors, hundreds of hours of additional review time have been spent by USGS personnel in attempts to correct and upgrade the documents.

Although calculations of the cost of that review are imprecise, conservative estimates by those involved are that several hundred thousand dollars have been added to the

$10 million already allocated to contractors.

Some private companies that have purchased "open file," or published, maps under the program have noted angrily that they could have compiled the same documents, containing more data for as little as half the cost.

OTHER SOURCES, both in BLM and private companies, have complained that all the information contained on some maps already is available in published form. For their part, the contractors defend the general purpose of the coal-reserve mapping program. "What we do is gather existing data," said Clarence Felix, chief geologist for AAA Engineering and Drafting, which is under contract to map 78 quadrangles. "Bringing all of that together and making other maps is useful to the Bureau of Land Management."

AAA President Ross Anderson said the contracts for the program will account for about 15 percent of his company's business this year.

B.B. RANDOLPH, an official of Dames and Moore, a large engineering consultant that has contracts to map 114 quadrangles at a total cost of about $1 million, said, "The contract isn't all that bad. It has some deficiencies."

Strong, detailed criticism of the program from within the USGS began more than three years ago, with personnel warning of pitfalls in the concept of the project and the duplication of existing published information.

Criticism voiced, but never fully answered includes:

- An April 1976 memo signed by 15 geologists urging that "maps should not be produced which are not justified by the available coal information." Since then, contractors have completed mapping quadrangles containing less than 1 percent of federal coal.

- A November 1978 memo noting that future contracts planned under the program included the Kaiparowits plateau in Utah, "an area already evaluated in detail to the complete satisfaction of the concerned BLM offices..."

- An April 1979 memo to the chief of the USGS conservation division noting the "absurdities" of mapping some federal coal whose location makes it inaccessible with current technology.

KASH ADMITTED that there are problems in the mapping program.

At the April 2 meeting in Denver with USGS personnel, Kash said, "When we contract out, we can expect to get less quality than when we do it (the work) in house."

But Kash argued later this month in a telephone interview that "tight deadlines" have necessitated parceling out the wok to several private companies.

However, several of these contracts have received lengthy time extensions.

Kash also said that if the maps were done by USGS staff, they would be more accurate, but would cost up to 10 times as much to produce.

USGS staff members reject this argument, saying that the time they spend correcting poorly drafted maps has pushed the cost of the maps well beyond the original dollar estimates.

At the April 2 meeting, Kash also admitted that part of the problem is the political aspect of the program; the fact that so much money already is committed to the project.

Glass said many difficulties could have been avoided by "scanning the literature and eliminating areas with little known coal."

What worries several USGS employees is that the survey apparently is using this program as a model for similarly mapping all leaseable minerals, such as phosphate and pot-ash.

As one staffer said, "I wouldn't be so upset, I'd stick with (the program) if this were going to be the end of it. But this isn't the end of it."

GEOLOGIC PLATES

Plates 1, 2, and 3 show snapshots of geologic maps in areas which will be familiar to tourists and residents alike. Geologic maps show each rock formation where it outcrops in a specific color. Each map covers approximately 54 square miles, or about 35,000 acres. Together, the maps shown cover about 162 square miles of the 1,300 square miles mapped by the author in Wyoming. For detailed geologic maps of a much larger scale, the bibliography listed the specific reference for maps which can be purchased from the U.S. Geological Survey. Those maps will include cross-sections through each map area as well as lithologic descriptions for all rock formation. If coal or phosphate resources exist in the map area, an economic evaluation will be given for those resources.

Plate I — **Geologic Map of the Rendezvous Peak quadrangle**

This map shows the geologic relationships in the southern part of the Teton Range. Teton Pass is only several hundred yards south of the map boundary. Phosphate is the mineral resource of economic interest.

Plate II — **Geologic Map of the Bull Creek quadrangle**

This map shows the geologic relationships along the Hoback River as well as the area along Granite Creek. The map covers the frontal zone of thrusting in the Idaho-Wyoming Overthrust Belt for this area. Coal and phosphate are the minerals of economic interest.

Plate III — **Geologic Map of the Camp Davis quadrangle**

This map shows the geologic relationships in the Hoback Junction area, where the Hoback River flows into the Snake River. Phosphate is the mineral of economic interest.

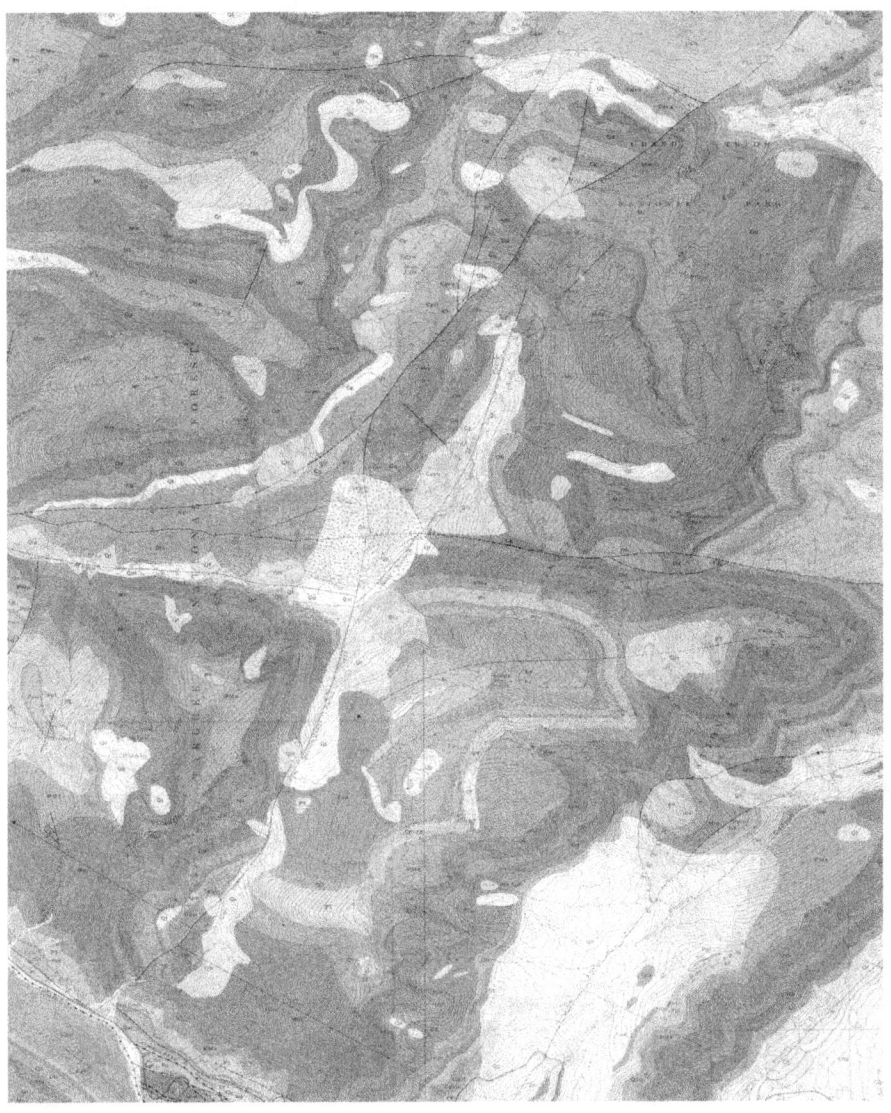

Plate I — Geologic Map of the Rendezvous Peak quadrangle

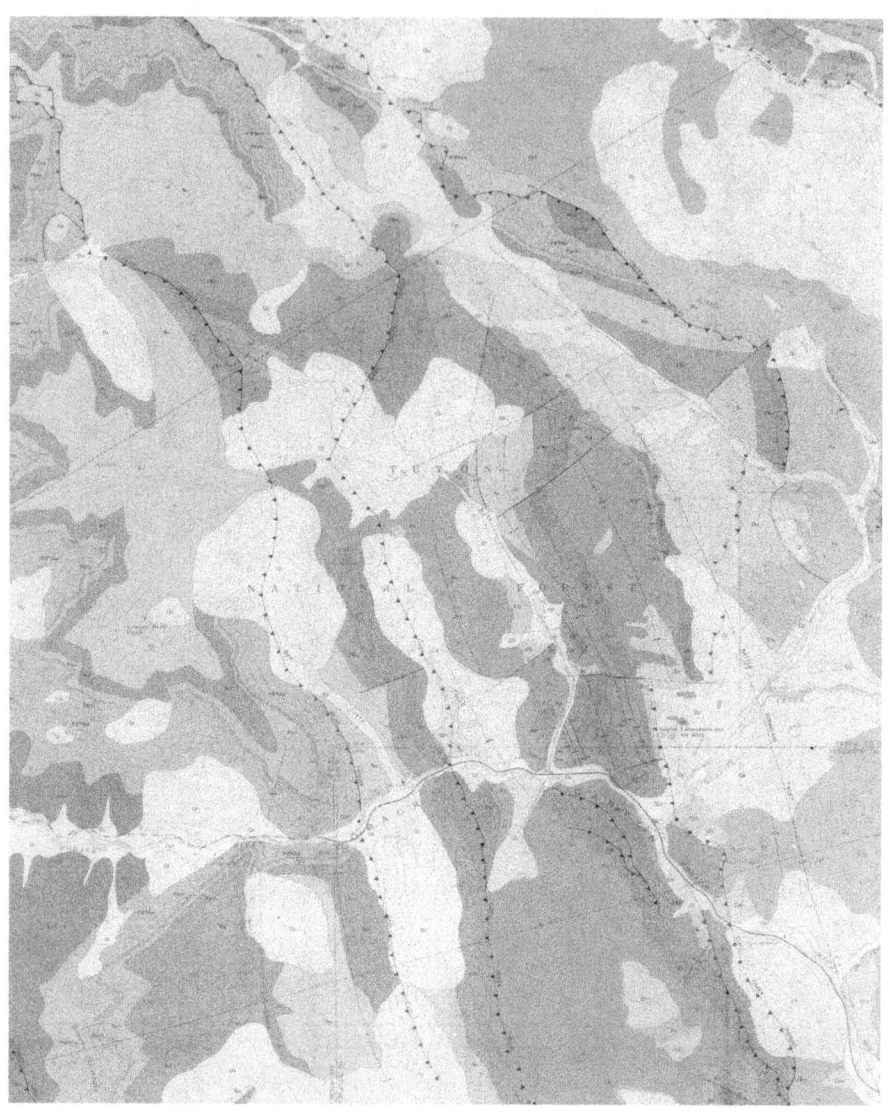

Plate II — Geologic Map of the Bull Creek quadrangle

Plate III — Geologic Map of the Camp Davis quadrangle

APPENDIX 2

The following fossil paper was published in 1968 in the journal of Micropaleontology. It is included here to give the reader a view of what a "fossil paper" looks like. This paper describes the best known occurrence of Lower Triassic arenaceous foraminifera from the Western Hemisphere. One new species is named for my wife, Linda, and another new species for my father. When a new fossil species is named in honor of a person, an "i" is added to a masculine name, e.g., Edward becomes Edwardi. An "e" is added to a feminine name, e.g., Linda becomes Lindae. Examples of letters received at that time asking for reprints are also included; these requests came from places as far away as Western Europe, India, and even the Republic of China.

Marvin L. Schroeder
United States Geological Survey
Denver, Colorado

Lower Triassic foraminifera from the Thaynes Formation in southeastern Idaho and western Wyoming

ABSTRACT

Approximately 600 specimens of arenaceous foraminifera have been recovered from insoluble residues of limestones from the Lower Triassic Thaynes Formation in southeastern Idaho and western Wyoming. Most of the ten species, four of which are new, belong to genera which are far-ranging in age. This is the second reported occurrence of foraminifera from the Lower Triassic in the Western Hemisphere.

INTRODUCTION

Foraminifera from Lower Triassic rocks are almost unknown in North America. The only occurrence previously reported is that described by Schell and Clark (1960), who have listed five silicified lagenid species, representing four genera, obtained from insoluble residues of limestones in northeastern Nevada. This report concerns 8 genera and 10 species of arenaceous foraminifera, 4 species of which are new, obtained from several thousand insoluble residues of limestone samples from the Lower Triassic Thaynes Formation in southeastern Idaho and western Wyoming. Several of the species are represented by only one or two specimens. Three species belong to the Ammodiscidae, two each to the Lituolidae and Textulariidae, and one each to the Astrorhizidae, Hormosinidae and Ataxophragmiidae. The specimens are better preserved and represent a more varied assemblage of Lower Triassic foraminifera than any known heretofore. The illustrations of the foraminifera were drawn by the author with the use of a camera lucida.

STRATIGRAPHY

The stratigraphic succession of the Lower Triassic strata in the Caribou Range in southeastern Idaho is given in text-figure 1. The fairly thick marine sequence of Triassic age lies along the eastern margin of a marine miogeosynclinal belt that existed here during Early Triassic time. The thickest and most complete section of Lower Triassic rocks in the miogeosyncline is in the area near Locality 7 (text-figure 2). According to Kummel (1954), "to the east, north, and south [of this area], the Dinwoody and Thaynes formations thin and intertongue with the red beds of the Chugwater, Woodside, and Ankareh formations."

The Thaynes Formation in the Caribou Range, about 960 feet thick, consists mainly of a blue-gray to greenish-gray, thin-bedded to medium-bedded, medium-grained, silty bioclastic limestone and interbedded calcareous sandstone and siltstone, with no red-bed intervals. However, nearly all sections of the Thaynes that the author has seen east of the Caribou Range do contain red-bed units, some intervals as much as several hundred feet thick. The distinct ammonoid zones recognized by Smith (1932) in the Thaynes of southeastern Idaho, in ascending order the *Meekoceras, Tirolites* and *Columbites* zones, are not present at Localities 1 and 8. Because the ammonoid zones are found only in the thicker geosynclinal sections to the south of this area, it is impossible to assign the arenaceous foraminifera to any one of the Scythian ammonoid zones.

254

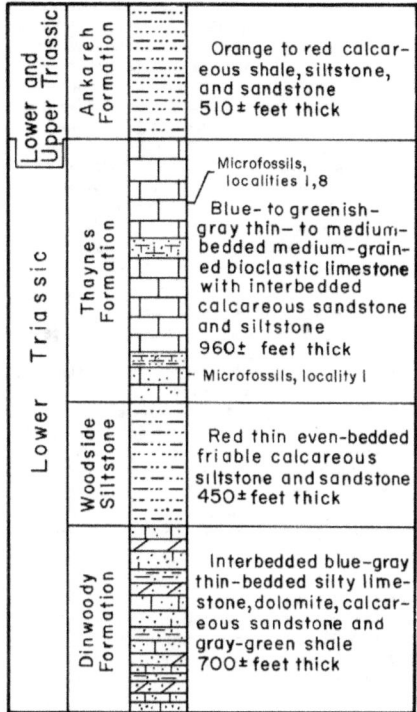

TEXT-FIGURE 1
Stratigraphic succession of the Lower Triassic in the Caribou
Range in southeastern Idaho.

The column contents (left to right) in the stratigraphic figure:

Lower and Upper Triassic	Ankareh Formation	Orange to red calcareous shale, siltstone, and sandstone 510± feet thick
Lower Triassic	Thaynes Formation	Microfossils, localities 1,8 Blue- to greenish-gray thin- to medium-bedded medium-grained bioclastic limestone with interbedded calcareous sandstone and siltstone 960± feet thick Microfossils, locality 1
	Woodside Siltstone	Red thin even-bedded friable calcareous siltstone and sandstone 450± feet thick
	Dinwoody Formation	Interbedded blue-gray thin-bedded silty limestone, dolomite, calcareous sandstone and gray-green shale 700± feet thick

TEXT-FIGURE 1
Stratigraphic succession of the Lower Triassic in the Caribou Range in southeastern Idaho.

COLLECTION OF SAMPLES

The Thaynes was extensively chip-sampled at 1-foot intervals at nine localities during a mapping program by the U. S. Geological Survey in this general area. All limestone samples were dissolved in weak (10 per cent) hydrochloric acid. When any evidence of microfossils was observed, additional limestone samples from the original collection were dissolved in hydrochloric acid and also in acetic acid to obtain any phosphatic forms that might be present. More samples were also collected from the intervals in which the microfossils were found.

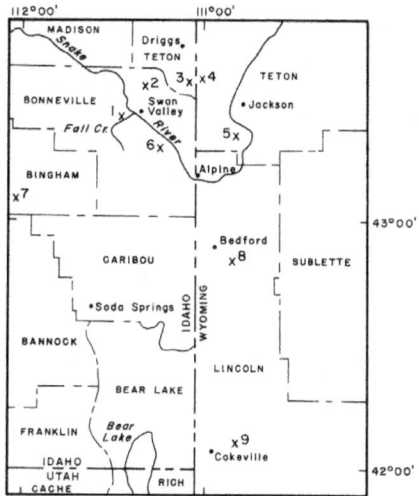

TEXT-FIGURE 2
Index map of localities sampled for foraminifera. Detailed descriptions of localities are listed in text.

The main collection of foraminifera was obtained from the upper 2 feet of a 6-foot, dark-gray, fine-grained carbonaceous limestone interval in the Caribou Range in southeastern Idaho along Fall Creek, about 260 feet from the top of the Thaynes. Associated with the foraminifera at this locality was a varied assemblage of other types of microfossils, consisting of conodonts, sponge spicules, shark teeth and brachiopod spines. Of these, the sponge spicules and shark teeth are relatively numerous and the conodonts relatively scarce — a sample containing only one conodont might have as many as a dozen specimens of the other types.

All attempts to obtain calcareous forms from the Fall Creek locality have been unsuccessful. Thin sections of the upper 2 feet of the limestone did not reveal any calcareous or arenaceous forms. At the interface between the dark-gray limestone and the overlying siltstone, siltstone samples were collected and broken down by boiling in the expectation that calcareous forms might be obtained, but no microfossils or fragmented fossils of any kind were found. Apparently, the upper 2-foot interval contains microfossils concentrated in thin bands rather than scattered throughout the limestone.

74

The specimens of *Hyperammina* in the collections were obtained about 70 feet above the base of the Thaynes at the Fall Creek locality in a very thin, dark-gray, fine-grained silty limestone bed. Several attempts have been made to relocate the limestone bed from which the original chip sample came, but all attempts have been unsuccessful.

The specimens of *Reophax* were obtained from a 15- to 20-foot, dark-gray to blue-gray silty limestone interval at Locality 8, in western Wyoming, approximately the same distance from the top of the Thaynes as the collection from Fall Creek, Locality 1. No other micro-fossils were found associated with the foraminifera at this locality. Although all the localities have been extensively sampled, only these two localities, 1 and 8, have yielded any foraminifera.

Detailed description of localities sampled for foraminifera:
1) Fall Creek locality, eastern Idaho, N½SE¼ sec. 18, T. 1 N., R. 43 E., Boise meridian.
2) Pine Creek locality, eastern Idaho, C SW¼ sec. 29, T. 3 N., R. 44 E., Boise meridian.
3) Pole Canyon locality, eastern Idaho, C NE¼ sec. 27, T. 3 N., R. 45 E., Boise meridian.
4) Nordwall Canyon locality NW¼SE¼ sec. 8, unsurveyed, T. 41 N., R. 118 W., sixth principal meridian.
5) Rock Creek locality, SE¼ sec. 8, unsurveyed, T. 39 N., R. 117 W., sixth principal meridian.
6) Calamity Point locality, NE¼SE¼ sec. 18, T. 1 S., R. 44 E., Boise meridian.
7) Ross Creek locality, SE¼ sec. 12, unsurveyed, T. 4 S., R. 37 E., Boise meridian.
8) Turnerville locality, western Wyoming, C S½ sec. 18, unsurveyed, T. 33 N., R. 117 W., sixth principal meridian.
9) Cokeville locality, S½ sec. 35, T. 25 N., R. 118 W., sixth principal meridian.

CHARACTER OF THE TRIASSIC FORAMINIFERA
The assemblage of foraminifera found in the Thaynes Formation differs from other described Triassic faunas in the United States. Four of the ten species belong to the superfamily Ammodiscacea, one of these in the family Astrorhizidae and three in the Ammodiscidae. The remaining six species belong to the superfamily Lituolacea, one in the Hormosinidae, two in the Lituolidae, two in the Textulariidae, and one in the Ataxophragmiidae. All five of the species described from the Lower Triassic in Nevada by Schell and Clark (1960) were silicified forms belonging to the family Nodosariidae. The only other Triassic fauna in North America with which one can make a comparison is that described by Tappan (1951) from the Upper Triassic of northern Alaska. Of the 26 species described, 12 belong to the Nodosariidae, 5 to the Polymorphinidae, 2 to the Lituolidae, 2 to the Trochamminidae, and 1 each to the Ammodiscidae, Ataxophragmiidae, Bolivinitidae, Spirillinidae and Discorbidae.

In the Triassic occurrences in Nevada and Alaska the Nodosariidae are predominant. No species of Nodosariidae were found in the Thaynes, but this would be expected since the specimens were recovered by dissolving limestones in acid and any calcareous forms would be lost in the process, unless replaced by siliceous or other insoluble material. No zonations can be established until more is known about the distribution of these species. Most of the species belong to simple arenaceous genera that, with the exception of *Verneuilinoides*, range throughout the post-Paleozoic, and some of the simpler forms, such as *Tolypammina* and *Hyperammina*, are known from the Ordovician and Silurian to the Recent.

COMPOSITION OF TESTS OF TRIASSIC FORAMINIFERA
It does not seem probable that secondary silicification has affected the original composition of the foraminiferal tests found in the Thaynes Formation because 1) most of the tests of the foraminifera are in a good state of preservation and 2) no representative of the Nodosariidae was recovered. Although the latter might be considered questionable evidence, it would seem logical that if any secondary silicification had occurred, it would have also included the calcareous tests of the Nodosariidae, since the Nodosariidae have dominated the foraminiferal assemblages previously described from Triassic strata.

ACKNOWLEDGMENTS
The author wishes to acknowledge the helpful criticisms of the original manuscript made by James F. Mello of the Branch of Paleontology and Stratigraphy, United States Geological Survey. Publication has been authorized by the Director of the United States Geological Survey.

SYSTEMATIC PALEONTOLOGY
All type forms and figured specimens are to be deposited in the U. S. National Museum, and their serial numbers are included in the plate explanations. Unless otherwise indicated, all forms were obtained from the upper part of the Thaynes at Locality 1. The classification followed in the text is that given by the *Treatise on Invertebrate Paleontology* for the "Foraminiferida".

Order FORAMINIFERIDA
Superfamily AMMODISCACEA
Family AMMODISCIDAE
Subfamily TOLYPAMMININAE
Genus AMMOVERTELLA Cushman, 1928

Ammovertella liassica Barnard
Plate 1, figures 1–2

Ammovertella liassica BARNARD, 1950, p. 354, text-fig. 1c.

75

256

Description: Test attached, with attached side flat and unattached side convex, consisting of a proloculus and a tunnel-shaped second chamber initially planispirally coiled with adjacent parts of tube used as wall, then extending out in one general direction of growth; extended tube increasing only slightly in diameter; whorls about four in number; wall composed of fine quartz silt, well-cemented; aperture at end of extended tube.

Dimensions: Specimen illustrated in figure 1, diameter of coil 0.28 to 0.35 mm., length of extended tube 0.42 mm.

Remarks: Barnard described this species from the Lower Jurassic of the Dorset coast, England, where he found it to be fairly common. Only two complete specimens were found in the Thaynes, but they fit Barnard's description and illustration perfectly. The direction of growth of the planispirally coiled portion of the test may be either clockwise or counterclockwise. Both types of coiling are illustrated here. The attached side (figure 1b) is partly crusted over, hence the discontinuity of the chamber in places.

Ammovertella inclusa ? (Cushman and Waters)
Plate 1, figure 4a–b

?*Psammophis inclusus* CUSHMAN and WATERS, 1927, p. 148, pl. 26, fig. 12.
?*Tolypammina inclusa* (Cushman and Waters). – GALLOWAY and RYNIKER, 1930, p. 11, pl. 1, figs. 12–13.
?*Ammovertella inclusa* (Cushman and Waters). – CUSHMAN and WATERS, 1930, p. 44, pl. 7, fig. 13.

Description: Test attached, incomplete with initial portion broken off, attached side flattened, unattached side convex, consisting of an elongate tunnel-shaped chamber that enlarges gradually, uses adjacent parts of the tube as a common wall and is flexed back and forth in a common plane with later stages extended over earlier portions of tube.

Dimensions: Diameter of test 0.35 to 0.40 mm., diameter of tube 0.07 to 0.10 mm.

Remarks: Only one specimen was found. The specific determination is questioned since the initial portion of the test is gone, but the portion of the specimen present is similar to the corresponding portion of *A. inclusa.* Cushman and Waters (1930) found this species in the South Bend Shale Member of the Graham Formation of the Cisco Group in central Texas. *Ammovertella elevata* Ireland is also similar in its general shape and size, but the tube of that species coils more extensively over the test, elevating it much higher than in *A. inclusa.* Ireland (1956) found this species to be present in many of the beds of Virgil age in Kansas.

Genus TOLYPAMMINA Rhumbler, 1895

Tolypammina sp.
Plate 1, figure 3

Description: Test incomplete, consisting of a tubular second chamber partially coiled, with the proloculus and initial portion broken off; tube increasing only slightly in diameter; wall fine to medium quartz silt, well-cemented; surface smooth to slightly rough; aperture at open end of tube.

Dimensions: Length of tube about 1.2 mm., greatest width of tube 0.13 mm.

Remarks: Only one specimen of *Tolypammina* was found in the Thaynes. Not enough of the specimen is present to define it at the specific level. However, its general dimensions and smoothness of test wall suggest that it is most similar to *Tolypammina polyverta* Ireland. Ireland (1956) found this form to be the most common of the tolypamminids in the rocks of the Virgil Series in Kansas.

Family HORMOSINIDAE
Genus REOPHAX Montfort, 1808

Reophax finleyi Schroeder, **n. sp.**
Plate 1, figures 5–9

Description: Test small, elongate, composed of a proloculus and as many as eight gradually expanding chambers; last chamber about as wide as high, only occasionally pyriform in shape; preceding chambers usually tapering to initial extremity, with greatest diameter usually in upper portion of chamber; early chambers with a greater diameter than height; chambers rarely aligned along a straight axis, greatest deviations usually occurring in earlier portions of test; length of test ranging from about 4.5 to 5.5 times the maximum diameter; proloculus and the following several chambers broken off in many specimens; aperture a simple rounded terminal opening in last chamber; test of medium-coarse quartz silt, with fine quartz grains occasionally incorporated into test; color of test white.

TABLE 1
Measurements of *Reophax finleyi* in mm.

Specimen (Pl. 1)	Length	Height, last chamber	Diameter, last chamber	Height, first chamber	Diameter, first chamber
Fig. 5	0.46	0.07	0.10	0.03	0.04
Fig. 6	0.51	0.10	0.09	0.03	0.04
Fig. 7	0.56	0.11	0.11	0.05	0.09
Fig. 8	0.76	0.14	0.13	0.03	0.06

Dimensions: See table 1.

Remarks: The diagnostic features of this new species are the slowly expanding chambers, producing a test that has a length of 4.5 to 5.5 times the maximum chamber diameter, the sharply incised sutures, the single simple aperture at the end of a regular final chamber, the medium-coarse grain silt test, the low early chambers, and the later chambers of approximately equal dimensions.

Both *Reophax northviewensis* Conkin and Conkin, 1964, and *Reophax minutissimus* Plummer, 1945, are somewhat similar to *R. finleyi* in the length of the test, number of chambers, and low early chambers, but differ from this new species in having pyriform chambers, an aperture at the end of a short neck extending from the last chamber, a greater diameter of the last chamber, and a length shorter in comparison with the maximum diameter.

This new species is well represented in the collections, more than 60 specimens being present. Most of the specimens, however, are only fairly well preserved. The holotype is one of the bigger specimens found. Measurements for the specimen illustrated by figure 9 on plate 1 are not given because it is an internal mold and the dimensions might be misleading. An indication of the thickness of the test wall may be obtained by comparing the mold to figure 8, which is about the same height. The pyriform final chamber of the specimen illustrated by figure 5 on plate 1 is rare in that the height/diameter ratio of the final chamber is usually about equal. Because the proloculus is missing on many specimens, it is impossible to determine its exact form, but, where observed, it generally seems to be round to slightly prolate.

This new species of *Reophax* is named in honor of E. A. Finley, who is Chief of the Branch of Mineral Classification, U. S. Geological Survey.

Family ASTRORHIZIDAE
Subfamily HIPPOCREPININAE
Genus HYPERAMMINA Brady, 1878

Hyperammina glabra ? Cushman and Waters
Plate 1, figures 10–11

?Hyperammina glabra CUSHMAN and WATERS, 1927, p. 146, pl. 26, fig. 1.

Description: Test free, elongate, consisting only of tapered tubular second chamber, incomplete, initial extremity broken off; wall of fine arenaceous material with much cement, with very fine quartz sand and euhedral quartz crystals occasionally incorporated into cement; aperture formed by open end of tube.

Dimensions: Specimen illustrated in figure 10, length of test 0.58 mm., width of tube 0.05 to 0.10 mm.; specimen illustrated in figure 11, length of tube 0.50 mm., width of tube 0.06 to 0.11 mm.

Remarks: Only eight fairly complete specimens were found, along with a lot of other fragmented material. Each of the eight specimens lacks a proloculus. Because the proloculi are missing, the specific determination is questioned, but these specimens otherwise fit fairly closely the description given by Cushman and Waters for *H. glabra.* Cushman and Waters (1927) found this species near the top of the Strawn and also (1930) in the South Bend Shale Member of the Graham Formation of the Cisco Group. Ireland (1956) found this species in many of the beds throughout the Virgil Series in Kansas.

This form was found about 70 feet above the base of the Thaynes along Fall Creek, Locality 1. Various attempts to find the original limestone bed from which the samples were obtained have been unsuccessful.

Superfamily LITUOLACEA
Family LITUOLIDAE
Subfamily LITUOLINAE
Genus AMMOBACULITES Cushman, 1910

Ammobaculites duncani Schroeder, **n. sp.**
Plate 1, figures 12–16

Description: Test elongate, slender, consisting of a proloculus and usually four (only seldom five) chambers planispirally coiled, followed by a maximum of eight rectilinear inflated chambers which increase in diameter only slightly, producing a terminal portion with essentially parallel sides; initial rectilinear chambers of microspheric form lower than those of megalospheric form, resulting in several additional chambers in tests of equal length; sutures distinct, slightly depressed; wall composed of fine-grained quartz silt, well-cemented, surface smooth to slightly rough; aperture circular, terminal in last chamber; color of test white to rusty gray.

TABLE 2
Measurements of *Ammobaculites duncani* in mm.

Specimen (Pl. 1)	Length	Diam., coil	Length, rect. part	Diam., last chamber	Height, last chamber	No. of coiled chambers	No. of rect. chambers
Fig. 12	0.65	0.14	0.53	0.10	0.12	4	6
Fig. 13	0.58	0.12	0.46	0.10	0.13	5	6
Fig. 14	0.63	0.13	0.52	0.11	0.12	4	5
Fig. 15	0.54	0.09	0.47	0.10	0.11	4	7
Fig. 16	0.31	0.11	0.22	0.09	0.09	4	3

Dimensions: See table 2.

77

Remarks: The diagnostic features of this new species are the four chambers in the planispiral coiled portion and as many as eight chambers in the rectilinear portion, the sutures sharply incised into the wall, the finely arenaceous wall, and the simple aperture at the end of the final chamber.

Ammobaculites parallelus Ireland from the Pennsylvanian of Kansas is similar to *A. duncani* in its long test and rectilinear series with essentially parallel sides, but it differs in having more chambers, five to six in the coil instead of usually four, and in having a test composed of medium to coarse quartz silt.

Ammobaculites inconspicuus Cushman and Waters from the Pennsylvanian of Texas is similar in the dimensions of the test, the finely arenaceous wall, and the number of chambers in the rectilinear series, but *A. duncani* differs in usually having four chambers in the coil and a rectilinear series in which the chambers are usually parallel.

Ammobaculites rectus (Brady), as described by Harlton (1928) from the Pennsylvanian of Texas, also has a long test with a smooth finish, but it has a gently tapering test and more than four chambers in the coil. *Ammobaculites fisheri* Crespin, 1953, from the Lower Cretaceous of Australia is similar in the well-defined chambers in the planispiral and uniserial portions of the test, and the small number of chambers, three to five, in the coil, but differs in usually having five chambers in the rectilinear portion, a subpyriform last chamber, a rough surface, and an aperture that varies in shape from circular to elliptical and is surrounded by a slight lip.

This form is the most abundant species of the foraminifera found in the Thaynes, more than 300 specimens being present in the collections. Of all the specimens, 98 per cent or more have four chambers in the coil, which is the dominant characteristic separating the species from others with an elongate rectilinear series. Some of the longer specimens show a tendency to be slightly twisted. This twisting is believed to happen during the final compaction of the sediment rather than to be an inherent characteristic of the species, since none of the shorter specimens show this characteristic. One of the remarkable features of the species is its selection of grain size for the wall. All of the specimens in the collections have used a very fine-grained quartz silt in the construction of the test.

This new species of *Ammobaculites* is named in honor of H. J. Duncan, formerly with the U. S. Geological Survey, who recently retired after 22 years of service as Chief of the Conservation Division.

Subfamily HAPLOPHRAGMOIDINAE
Genus TROCHAMMINOIDES Cushman, 1910

***Trochamminoides* sp.**
Plate 1, figure 17a–b

Description: Test free, tiny, planispiral, not involute, consisting of about two whorls ; about seven chambers in last whorl ; chambers not globular but enlarging regularly from proloculus to last chamber ; sutures sharp, only slightly depressed ; wall finely arenaceous and smoothly finished ; aperture simple, at end of the last chamber.

Dimensions: Greatest diameter of specimen represented in figure 17 0.33 mm., least diameter 0.25 mm., greatest thickness 0.13 mm.

Remarks: Only two specimens were found, but both are well preserved and well developed. *Trochamminoides vertens* Tappan, 1957, from the Upper Triassic of Alaska is similar to the two specimens found in the Thaynes in general size, but differs in having more globular and inflated chambers and nine chambers to a whorl. It is not believed wise to erect a new species on the basis of two specimens, and designation of a new species will be delayed until more specimens can be obtained.

Family TEXTULARIIDAE
Subfamily TEXTULARIINAE
Genus BIGENERINA d'Orbigny, 1826

***Bigenerina lindae* Schroder, n. sp.**
Plate 1, figures 18–23

Description: Test small, elongate ; biserial portion fairly blunt to gradually tapered, consisting of seven to eight chambers ; rectilinear series composed of three to four inflated chambers, with first chamber wedge-shaped and sharply set off at an angle from biserial portion, later chambers usually having sinuous or winding growth in a plane identical with a plane determined by the length and the width of the biserial portion ; chambers of rectilinear series usually of equal width and height but early chambers may occasionally be depressed ; sutures well defined and depressed in recti-linear series but only fairly well defined and depressed in biserial portion ; wall of fine to coarse quartz silt, well-cemented ; aperture round and terminal at end of a low protuberance on last chamber.

Dimensions: See table 3.

Remarks: The diagnostic features of this new species are the sinuous growth in the uniserial series in a plane identical with that of the biserial part, the inflated uni-serial chambers, the circular aperture at the end of a small protuberance on the last chamber, the well-

78

TABLE 3
Measurements of *Bigenerina lindae* in mm.

Specimen (Pl. 1)	Length	No. of biserial chambers	No. of uniserial chambers	Diameter, last chamber	Width, biserial part
Fig. 18	0.40	8	4	0.10	0.10
Fig. 19	0.41	7	4	0.13	0.10
Fig. 20	0.42	7	3	0.13	0.13
Fig. 21	0.32	8	3	0.09	0.10
Fig. 22	0.30	7	2	0.10	0.11
Fig. 23	0.28	8	2	0.11	0.11

defined break between the biserial and uniserial parts, and the coarseness of the test.

Bigenerina burri Finlay, 1947, from the Paleocene of New Zealand is similar to *B. lindae* in the coarseness of its test, the number of chambers in the biserial portion, and the sinuous growth of the early chambers in the uniserial portion. *B. lindae* differs, however, in usually having inflated uniserial chambers, a circular aperture at the end of a small protuberance on the last chamber, sinuous growth of the uniserial series in a plane identical with that of the biserial part, and a well-defined break between the biserial and uniserial part. *B. burri* shows a transition between the biserial and uniserial parts, the first two chambers in the uniserial portion being loosely biserial before being followed by one to two normal uniserial chambers.

Bigenerina lindae is well represented by more than 60 specimens in the collection. It is believed that the figured specimens show most of the range of variation in the species. The specimen illustrated in figure 21 on plate 1 is unique in that it is the only specimen found in which the growth of the chambers in the uniserial portion of the test is in a plane perpendicular to the greatest width of the biserial portion. This species has probably been the hardest to define of any encountered in the Thaynes Formation because of the coarse quartz silt frequently found in the uniserial portion of the test, masking the sinuous growth of many of the specimens.

Bigenerina sp. cf. **B. perexigua** Plummer
Plate 1, figure 24

Cf. *Bigenerina perexigua* PLUMMER, 1945, p. 243, pl. 16, figs. 19-20.

Description: Test small, elongate; biserial portion tapered, consisting of 8 to 10 chambers with early chambers poorly defined; rectilinear portion straight, composed of four chambers of uniform width; not much change in width of test between rectilinear series and biserial portion; wall composed of medium-grained quartz silt, well-cemented, having a smooth appearance; sutures sharply distinct but not greatly

depressed; aperture central and terminal on last chamber.

Dimensions: Length of test 0.38 mm., width of rectilinear series 0.10 mm., length of biserial portion 0.15 mm.

Remarks: Only one well-preserved specimen is in the collections, but it is very similar to the description given by Mrs. Plummer (1945) for forms that she found in the Pennsylvanian rocks of central and northern Texas. Ireland (1956) also found this form in some of the limestones of the Virgil Series in Kansas.

Family ATAXOPHRAGMIIDAE
Genus VERNEUILINOIDES Loeblich and Tappan, 1949

Verneuilinoides edwardi Schroeder, **n. sp.**
Plate 1, figures 25-26

Description: Test free, small, triserial, generally subcircular in transverse section but sometimes subtriangular; chambers tightly coiled, not globular, enlarging slowly; sutures slightly depressed, well defined only between last several chambers; wall of fine to medium-grained quartz silt, well-cemented; surface smooth to slightly rough; aperture a very low arch at base of last chamber.

Dimensions: Specimen illustrated in figure 25, height 0.25 mm., greatest width 0.14 mm.; specimen illustrated in figure 26, height 0.25 mm., greatest width 0.13 mm.

Remarks: The diagnostic features of this new species are the small size, the low apertural arch, the finely arenaceous wall, and the slowly enlarging chambers resulting in an appressed test.

More than 70 specimens of *V. edwardi* are in the collections, and this new species is quite different from all of the other described species of *Verneuilinoides*. The range of variation within the species is small. Almost invariably the height of the specimens is about 0.25 mm., varying no more than 0.02 mm. Occasionally a specimen will tend to have its sides close to parallel, as illustrated by figure 26. None of the specimens in the collections has any coarse-size material in the wall, which is composed of a very fine-grained quartz silt.

This is the first reported occurrence of *Verneuilinoides* in rocks of Triassic age, the *Treatise of Invertebrate Paleontology* listing the range of the genus as Jurassic to Cretaceous. Harlan Bergquist of the U. S. Geological Survey (written communication, 1966) has suggested that the specimens may be the triserial stage of a *Gaudryina*, which is the only genus of the subfamily Verneuilininae that the *Treatise* lists as occurring in rocks older than Jurassic. This, however, seems unlikely. Of the 70 or more specimens available in the collections, all appear triserial throughout and show no tendency to develop a biserial portion. If these were only the triserial stages of a *Gaudryina*, it would indicate either 1) that the population was killed off in an immature stage or 2) that the environment was too harsh for full development. However, the other foraminifera found in this same 2-foot zone in the Thaynes have well-developed populations with no sign of living in a harsh environment which might have retarded their growth or prevented full maturity. Consequently, the author regards these specimens as belonging to the genus *Verneuilinoides* rather than as being an initial stage of *Gaudryina*.

The range of the genus may actually extend back into the Paleozoic. Ireland (1956) described a new species of *Verneuilina* from the Virgil Series of the Pennsylvanian in Kansas. *Verneuilina* and *Verneuilinoides* are in the same subfamily and are very closely related, both being triserial and having an apertural arch at the base of the final chamber. *Verneuilinoides* differs from *Verneuilina* in having more inflated chambers and being rounded in cross section. Ireland's illustrations seem to more closely fit the description of *Verneuilinoides* than that of *Verneuilina*. If subsequent study shows that this species belongs to the genus *Verneuilinoides*, the range of *Verneuilinoides* will be extended from the Pennsylvanian to the Cretaceous. The study of Paleozoic foraminifera has been increasing in the last 10 years, and doubtless the range of more genera will be extended back into the Paleozoic as more information is obtained.

PLATE 1

All figures ×80

1–2 *Ammovertella liassica* Barnard
 1, USNM 643516, specimen showing clockwise coiling of planispiral portion; a, unattached side; b, attached side; 2, USNM 643517, specimen viewed from unattached side, showing counterclockwise coiling of planispiral portion, as well as partial overlap of tube over last whorl.

3 *Tolypammina* sp.
 USNM 643518, incomplete specimen.

4 *Ammovertella inclusa ?* (Cushman and Waters)
 USNM 643519, specimen showing meandering nature of tubular chamber; a, unattached side; b, attached side.

5–9 *Reophax finleyi*, Schroeder, n. sp.
 5, paratype, USNM 643520, showing coarseness of test, subpyriform last chamber, and contrast with other specimens; 6, paratype, USNM 643521, showing some tiny quartz crystals incorporated in the test; 7, paratype, USNM 643522, showing poorly preserved nature of test, fairly typical of many of the specimens found; 8, holotype, USNM 643523, showing slowly enlarging chambers and sutures sharply but not deeply incised into test; 9, USNM 643524, internal mold.

10–11 *Hyperammina glabra ?* Cushman and Waters
 10, USNM 643525, specimen showing slowly enlarging tubular second chamber; 11, USNM 643526, specimen showing sudden but uncommon enlargement of test.

12–16 *Ammobaculites duncani* Schroeder, n. sp.
 12, holotype, megalospheric form, USNM 643527, showing four chambers in coil, smoothness of test, and rectilinear series with chambers having parallel sides; 13, paratype, USNM 643528, showing un- common feature of five chambers in coil; 14, paratype, USNM 643529, showing slight deviation in suture line from the horizontal; 15, paratype, microspheric form, USNM 643530, showing depressed nature of some chambers, resulting in a greater number of chambers compared with a megalospheric form of equal length; 16, paratype, USNM 643531, immature specimen.

17 *Trochamminoides* sp.
 Holotype, USNM 643532, showing gradually enlarging chambers, sutures sharply but not deeply incised into test, and finely arenaceous test wall; a, side view; b, apertural view.

18–23 *Bigenerina lindae* Schroeder, n. sp.
 18, holotype, USNM 643533, and 19–20, paratypes, USNM 643534 and 634545, showing inflated wedge-shaped uniserial chambers having sinuous growth in plane identical with that of greatest length and greatest width of biserial portion, and circular aperture at end of small protuberance of last chamber; 21, paratype, USNM 643536, showing unique specimen having rectilinear series in a plane perpendicular to that of greatest length and greatest width of biserial portion; 22–23, paratypes, USNM 643537 and 643538, immature specimens.

24 *Bigenerina* sp. cf. *B. perexigua* Plummer
 USNM 643539, specimen showing uniserial chambers of equal width, sutures sharply incised in the test wall, and lack of restriction between the biserial and uniserial portions.

25–26 *Verneuilinoides edwardi* Schroeder, n. sp.
 25, holotype, USNM 643540, showing small size, slowly enlarging chambers, appressed test, and very low apertural arch; a, side view; b, apertural view; 26, paratype, USNM 643541, showing almost parallel sides, a rare feature in this species.

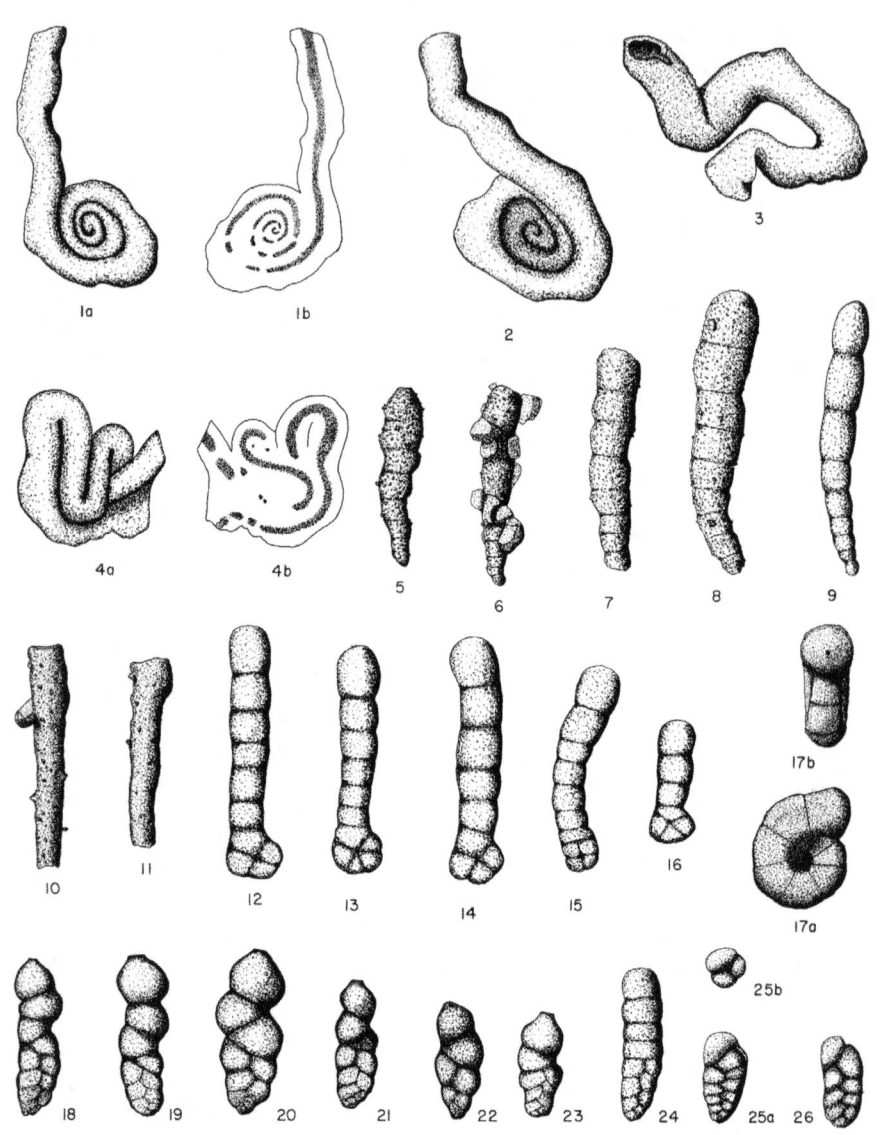

262

Budapest, 1. 1o. 197o.

Dear Sir!

I beg your pardon, that I disturb you with my letter
as an alien person. I'm a Hungarian geologist. I have graduated three years ago in the University of Budapest. Since
that I deal with Triassic microbiofacies and Triassic Fora-
minifers.

Since I need the up to date results in this field in
my work.

I ask you to send me reprints of your publications still
available, if it is possible.

I beg your pardon again because of my request, and I
say thank for your trouble.

Your faithfully:

Bérci-Makk, Anikó

My adress: Bérczi-Makk, Anikó
               Budapest, XI.
               Vasut utca 1/a. I. 4.
               Hungary

Typical letter I received after my fossil paper was published in 1968.

UNIVERSITÉ DE GENÈVE

INSTITUT DE GÉOLOGIE
ET DE PALÉONTOLOGIE
LABORATOIRE DE PALÉONTOLOGIE

---

11 b, rue des Maraîchers
1211 Genève 4 - Suisse
Tél. 24 12 00 & 24 12 09
Dr.R.Wernli

Pour les grandes villes
indiquez toujours
le numéro postal
→ ▸ ▸ du quartier ◂ ◂

1329

Dr.Marvin L.SCHROEDER

United States Geological
Survey

Denver
COLORADO   U.S.A.

Various envelopes showing requests from different countries.

This request for a reprint of my fossil paper was from the People's Republic of China

LIST OF GEOLOGISTS WHO WORKED IN MAPPING PROGRAM

*Albee, H.

Barnum, B.

*Barclay, C.

*Bowers, W.

Brownfield, M.

*Cole, T.

Cullins, H.

*Cressman, E.

Dickinson, R.

Dyni, J.

Eager, P.

Ege, J.

Ellis, G.

*Fraser, G.

Gaskill, D.

*Gere, W.

Godwin, L.

*Hyden, H.

Hettinger, R.

Hoover, D.

*Horn, G.

Izett, G.

*Jobin, D.

Johnson, E.

*Laraway, W.

Law, B.

Madden, D.

*Mullens, T.

Nutt, C.

Pampayean, E.

Peterson, F.

Reheiss, M.

Rohrer, W.

*Rioux, R.

*Schell, E.

Schroeder, M.

*Soister, P.

*Staatz, M.

Stevens, V.

*Waldrup, H.

*Wanek, A.

*Zeller, H.

*Deceased